CRTD-15-1

Selected Case Studies in the Use of

TOLERANCE AND DEVIATION INFORMATION

During Design of Representative Industrial Products

prepared by
VIJAY A. TIPNIS
SYNERGY INTERNATIONAL, INC.
under the auspices of the
ASME BOARD ON RESEARCH

for the
U.S. NATIONAL SCIENCE FOUNDATION
DESIGN AND MANUFACTURING SYSTEMS DIVISION
under
NSF GRANT DDM 90-00383

REPORT COMMITTEE
VIJAY A. TIPNIS, *Chairman*
H. BARRY BEBB — BARRY BEBB & ASSOCIATES (formerly with Xerox Corp.)
MIKEL J. HARRY — MOTOROLA
LOUIS LAVALLEE — XEROX CORP.
JEAN M. PARKS — XEROX CORP.
MADHAV S. PHADKE — PHADKE ASSOCIATES (formerly with AT&T Bell Labs)
LARRY R. SMITH — FORD MOTOR COMPANY

THE AMERICAN SOCIETY OF MECHANICAL ENGINEERS
United Engineering Center • 345 East 47th Street • New York, N.Y. 10017

This report is based on a workshop funded by NSF Grant DDM-9000383 to the ASME Board on Research and Technical Development.
NSF Program Director, Tony C. Woo.

Opinions, findings, and conclusions or recommendations expressed in this report are those of the committee and do not necessarily reflect the views of the National Science Foundation or ASME.

Statement from By-Laws: The Society shall not be responsible for statements or opinions advanced in papers . . . or printed in its publications (7.1.3)

ISBN No. 0-7918-0666-9

Library of Congress
Catalog Number 92-52849

PREFACE

These cases demonstrate how some of the leading U. S. Corporations have adopted and successfully implemented methods to make their products and processes robust through selection and control of appropriate tolerances and deviations. Hence, the case studies presented in this report should be of interest to industry practitioners. The report should also be of interest to the tolerance research community because it shows how some of the leading corporations in the U. S. mechanical parts industry are dealing with the issues of tolerances and deviations despite a number of major research gaps that remain un-addressed. It should interest the standards community because the cases show that the traditional mechanical tolerance standards need to incorporate some of the recent advances implemented in industry. And finally, it should be of interest to those involved in training engineers in industry and academe because it shows what should be taught and how extensive the training has to be to successfully adopt and implement robust design practices in industry.

The task of gathering materials for cases, obtaining permission for reviewing the drafts through several iterations would not have been possible without excellent cooperation from the managers and engineers of the participating companies. My sincere thanks, therefore, go to them as well as the authors themselves. ASME New York Art Director Carl MacIver gladly provided assistance in the layout of figures and tables in this document. Working with Norma Johnston and Ray Art of ASME, Center for Research and Technology Development, Washington D. C., has been a pleasure as always.

I acknowledged with gratitude, NSF sponsorship of this project which continued from the International Tolerancing Workshop under then NSF project Manager, Tony Woo.

Vijay A. Tipnis
Synergy International, Inc.
Atlanta, GA 30328 U. S. A.
Phone/Fax: (404) 851-0134

TABLE OF CONTENTS

INTRODUCTION

The need for this investigation was identified during the NSF workshop on Research Opportunities in Mechanical Tolerancing (Tipnis et al., 1989). Although mechanical tolerances are an important element for ensuring that the industrial products satisfy functional requirements, the 1988 ASME/NSF Tolerancing Workshop participants could not present any documented case studies as to why the specific tolerances were chosen and how these tolerances were achieved throughout the manufacturing and other life cycle phases of the product. In this investigation, the objective was to document case studies on how selected corporations design and implement products and processes when tolerances are important. Through these case studies, an identification of the important issues where generic research opportunities exist in representation of tolerances from the mission requirements through the various life cycle phases of design, process planning, production, quality control/inspection, and field application was accomplished.

During the course of this investigation, the following organizations were contacted (to some of these organizations, site visits were made) to determine which cases were most relevant to the investigation. Those highlighted were chosen for case studies.

(1) Power transmissions for helicopters, tanks/trucks, automobiles, and ships.

Bell Helicopter Co., Dallas, TX
GM Allison Transmissions, Indianapolis, IN
Ford Transmission Design Center, Detroit, MI
Westinghouse Corp., Marine Division, Sunnyvale, CA
Westinghouse Corp., Nuclear Division, Pittsburgh, PA
Garrett Corporation, Phoenix, AZ

(2) Electronics packaging and high volume plastics and stampings.

Motorola Government Electronic Division, Phoenix, AZ
Eastman Kodak Co., Rochester, NY
Xerox Corp., Webster, NY
IBM Corp., Endicott, NY
AT&T Bell Laboratories

(3) Electro-Optical-Mechanical systems and products.

Eastman Kodak Co., Rochester, NY
Xerox Corp., Webster, NY
Lawrence Livermore Nuclear Laboratories, Livermore, CA

The case studies were chosen on the basis of how well the methodologies and practices were implemented and how extensive the organizational commitment was to these methodologies and practices. By and large the case studies, therefore, represent acceptance of the methodologies and in most instances, wide spread use of the methodologies adopted by the companies.

TOLERANCE AND DEVIATION INFORMATION

(A commentary on the Standards, Research, and Implementation Issues and on the Case Studies on Robust Design Practices)

Vijay A. Tipnis*

SUMMARY

Specification, realization and verification (inspection) of tolerances and deviations are crucial engineering activities for achieving the required functional quality targets. The standards for specification of tolerances and deviations have evolved from empirical engineering practices without a sound mathematical foundation. Consequently, ambiguities and inconsistencies of these standards are known to create interpretation and verification problems. Furthermore, verification is facing the challenge of coping with lack of consistency in measurements obtained from different Coordinate Measurement Systems to which most industrial practice have switched from the surface plate and hard gages. Despite considerable attention from standards, research, and practitioner communities, progress is slow due to many challenging conceptual, mathematical, and implementation issues.

This investigation involved identifying and documenting how leading U. S. Corporations are achieving desired quality targets despite these limitations:

First, the tightening of the vintage statistical quality and process control methods has been achieved, in some cases, to 3.4 ppm from the commonly accepted 66,803 ppm opportunities for nonconformities.

Secondly, recognizing the fact that current specification are incapable of fully describing the quality requirements of real physical objects and the fact that conformance to specification does not guarantee equally satisfying quality, robust design methods for parameter and tolerance design introduced first in Japan are being successfully implemented as demonstrated by the case studies.

Thirdly, the setting of specific functional tolerance and deviation targets has evolved into a systematic method for listening to the customer's needs and for benchmarking against competitors; originating in Japan, this method has been adopted successfully by some leading U. S. corporations.

Finally, attempts towards a workable theory and methodology for modeling tolerances and deviations stochastically from the customer satisfiers/dissatisfiers to the targets for individual part tolerances and deviations appear to start paying off.

These case studies are not a result of isolated instances. They represent the results of massive training and implementation in these leading corporations. It is hard to envision adoption on this massive scale without a strong top management commitment and a significant change in the product development process itself as indeed has been reported in one case study.

Observations on the research needs and implementation challenges are presented for critique and debate.

---

* President, Synergy International, Inc., Atlanta, GA

INTRODUCTION

Despite the crucial importance of specification, realization, and verification (inspection) of tolerances and deviations for achieving the required functional quality targets, serious challenges exist as concluded by the International Workshop on Mechanical Tolerancing held under the auspices of ASME/NSF (Tipnis, et al., 1989). The workshop provided an impartial forum for discussion of issues vital to the entire international community. During the workshop, a sense of recognition evolved that the current standards indeed need a sounder mathematical foundation. Merely insisting on adoption of the current standards and more education about the standards will not overcome their fundamental deficiencies. Also, computer representation of mechanical tolerances as labels in a computer database was recognized to be inadequate without the underlying mathematical foundation. Of the several inadequacies of the current standards, theories and practices, the issue of how the lack of consistent algorithms for Coordinate Measuring Systems (CMS; CMMs are a subset) is posing a problem for the design, manufacturing, and quality control communities worldwide, drew the most attention. The state of inability to unambiguously convey mechanical tolerancing information humbled most participants.

The serious gaps and research challenges revealed through this workshop deserve attention from academics, standards communities, and practitioners involved in design, process planning, manufacturing, inspection, prototype testing, and maintenance; indeed, the entire life cycle functions. The safety, reliability, and product performance expectations of engineered products cannot be assured without a clear specification and realization of the allowable deviations in the geometry, dimensions, form, and surface topography. Material, subsurface damage, and other physical and chemical properties important to the function of a part or assembly have been outside the scope of mechanical tolerancing standards. These cannot be left out of consideration if the entire product life cycle issues are to be addressed. Also, non-rigid parts such as seals, O rings, gaskets, and compliant parts such as plastics must be included in the specifications.

Although mechanical tolerances are an important element for ensuring that the industrial products satisfy functional requirements, workshop participants could not present any documented case studies as to why the specific tolerances were chosen and how these tolerances were achieved throughout the manufacturing and other life cycle phases of the product.

The objective of this investigation was to document case studies on how selected corporations design and implement products and processes when tolerances and deviations are important. Through these case studies, an identification of the important issues where generic research opportunities exist in the representation of tolerances from the mission requirements through the various life cycle phases of design, process planning, production, quality control/inspection, and field application was accomplished.

During the course of this investigation, the following organizations were contacted (to some of these organizations, site visits were made) to determine which cases were most relevant to the investigation. Those **highlighted** were chosen for case studies.

(1) Power transmissions for helicopters, tanks/trucks, automobiles, and ships.

Bell Helicopter Co., Dallas, TX
GM Allison Transmissions, Indianapolis, IN
Ford Transmission Design Center, Detroit, MI
Westinghouse Corp., Marine Division, Sunnyvale, CA

Westinghouse Corp., Nuclear Division, Pittsburgh, PA
Garrett Corporation, Phoenix, AZ

(2) *Electronics packaging and high volume plastics and stampings.*
Motorola Government Electronic Division, Phoenix, AZ
Eastman Kodak Co., Rochester, NY
Xerox Corp., Webster, NY
IBM Corp., Endicott, NY
AT&T Bell Laboratories

(3) *Electro-Optical-Mechanical systems and products.*
Eastman Kodak Co., Rochester, NY
Xerox Corp., Webster, NY
Lawrence Livermore Nuclear Laboratories, Livermore, CA

The case studies were chosen on the basis of how well the methodologies and practices were implemented and how extensive the organizational commitment was to these methodologies and practices. By and large, the case studies therefore represent acceptance of the methodologies and in most instances, wide spread usage of the methodologies adopted by the companies.

In this paper I have presented a commentary on the standards, research, and implementation issues in the case studies as a starting point for critique and debate on what should be done next.

EVOLUTION OF MECHANICAL TOLERANCING

The evolution of mechanical tolerancing can best be understood from the perspective of the evolution of product design and manufacturing cycle. Starting with the pre-industrial revolution till the modern era of automation, the product design and manufacturing cycle have evolved through distinct phases (Voelker, 1988): (1) *No explicit specifications:* Artisan/builder and customer were present throughout the process; (2) *Explicit designs prepared by the artisans:* Crude models or sketches were prepared by artisans prior to construction; (3) *Artisan/designer became distinct from artisan/builder:* As products became more elaborate, customer specifications evolved into contractual descriptions and specifications; (4) *Adaptive serial assembly:* A single artisan or a team of artisans built a whole product from start to finish, a craft method used for multiple copies, the parts were manufactured to fit the evolving assembly by "file and fit" method; (5) *Artisans replaced by specialist builder and assembler:* Gage-based interchangeability replaced the versatile artisan by specialists/builders of standard parts and assemblers, evolving into an intricately hierarchical series of activities from parts fabrication to final assembly. As Voelker correctly points out, the contemporary product design/manufacturing cycle is a continuation of this specialization.

Design representation, especially technical graphics and drafting conventions were precursors to the evolution of tolerancing practice. Voelker (refer to his 1988 paper for details) has shown how the 1920-vintage drawings did not show explicit tolerances but specified manufacturing process, while the 1950's practice used +/- tolerance on the nominal values, and the 1970's drawings show tolerancing with full attributes of modern true position tolerancing conventions. Thus, mechanical tolerancing, as practiced, is based on agreed upon convention described in industry adopted standards such as ANSI, ISO, other national, military or industry standards. Tolerancing conventions are attached to manually or computer drafted drawings in the form of numerical values, symbols, and notes that must be interpreted to understand the implied meaning. Despite the well-intentioned and diligent efforts of the constantly active standards community, the tolerancing convention remains subject to misinterpretation because there is no mathematical theory to assure unambiguity, uniqueness, consistency, or completeness within a toleranced drawing. Therefore, it should not come as a surprise to most that engineering drawings (views and isometrics) drawn manually or by wire-frame computer graphics (drafting) representations do not guarantee that the object represented is unambiguous or physically realizable. Solid modeling schemes have overcome this shortcoming. However, like the prior manual and computer drafting schemes, only nominal solids are modeled; tolerances are attached as labels.

Ambiguities in Mechanical Tolerancing Conventions

The tolerancing conventions in use today such as ANSI Y14.5, ISO, and others are so-called geometric tolerancing methods which define various geometric features such as straightness, flatness, circularity (roundness), position, perpendicularity, parallelism, angularity, coaxiality and others as bounded by regions of similar perfect geometry, offset from the nominal surface feature. Ambiguities in these and other aspects of ANSI Y14.5 as well as those of ISO were a significant topic of discussion at the last ASME/NSF International Workshop (Tipnis et al., 1989). Following one of the key recommendations of the workshop, ASME Codes and Standards Y14.5 Committee, under the sponsorship of NSF has undertaken an investigation to address the ambiguities and propose a more rigorous foundation. The present work is pursued under ANSI Y14.5.1 subcommittee with the goal of providing a firmer basis so as to reduce the risk of misinterpretation and ambiguities through definitions that are representation-free and non-computational and hence minimize the dependencies on specific representations, coordinate systems, and choice of origin. The key findings to date are (Srinivas, 1991):

(1) The standards suffer from: (a) *Syntactic* limitation: Symbolic annotations are based on two-dimensional projected views of part and assemblies, and (2) *Semantic* limitation: A lack of rigor in interpreting meaning behind the symbols. Another basic problem is the lack of a formal definition of 'fitting' verification data to the toleranced features. For example, the semantics of *size* tolerance is based on the use of two point measurements by a micrometer or a caliper. This poses serious limitations to the use of improved inspection technologies. How to maintain semantics free from any particular technique is a major concern. Progress appears possible by using theoretical inspection procedures (Requicha, 1983, 1984) which allow perfect-form and imperfect-form semantics (which does not depend on 'fitting').

(2) Notion of 'fitting' is invoked directly or indirectly in current standards: (a) *Datums:* In ANSI Y14.5 as well as in ISO, the definitions of datum refer to surface plate or hard gage; CMS use requires a mathematical equivalent in the form of an algorithm to fit datums; (b) *Size:* ANSI defines size between two parallel planes as actual or measured size rooted in the Taylor principle; ISO defines it as the local actual size based on the notion of cross section (The principal difference between the two is that while ANSI Y14.5 follows the Rule 1 (Taylor Principle) in which a boundary of perfect form at Maximum Material Condition (MMC) controls the variations in form as well as of individual features, the ISO recognizes the Independence Principle in which the dimensional and form tolerances must be met independently. The former means that a shaft, for example, be contained in an envelope of perfect form and shall have neither straightness nor circularity deviations; the latter controls dimensional tolerances separate from form deviations (Henzhold, 1991); (c) *Position:* Location as well as orientation are defined with respect to a datum referenced to the position of a particular geometric feature specified.

(3) The problem of methods divergence in CMMs (Feng and Hopp, 1991): Different measurement techniques, using the same data analysis algorithms, lead to different measurement results; also, different data algorithms produce different results using the same measurement techniques. This situation has pointed out fundamental flaws in the definitions of geometric dimensions and tolerances, data analysis and fitting algorithms, surface sampling techniques, and in CMM software. Several different efforts are underway as mentioned in this paper and elsewhere (for example, ANSI B89 Committee and CIRP STC D round robin for CMM's). However, a concern has arose that computationally attractive algorithms may not be consistent with the sound metrological principles. Indeed, the function of the feature must be addressed, not just geometry, as pointed out later in this paper.

Towards a Viable Theory of Mechanical Tolerancing

The need for a viable theory which can overcome the shortcomings of the current tolerancing conventions is sought by many researchers (mentioned by Tipnis, et al., 1988, Feng and Hopp, 1991, among others); however, to my knowledge, a comparative analysis of the strengths and weaknesses of the different approaches does not exit as yet. A viable theory must address the issues of (a) representation for parts and assemblies, including algorithms for tolerance analysis and synthesis, (b) verification problems pointed out above, (c) representational requirements for design, manufacturing, inspection as well as life cycle concerns such as safety and reliability, and (d) software issues of making the tolerancing capability available to a user without the mathematical complexity. These issues were debated at the ASME/NSF International Workshop on Tolerancing in 1988 (Tipnis, et al., 1988) which produced recommendations for research. To illustrate how design, verification and manufacturing issues could be treated, one approach is briefly reviewed below because it has a bearing on the industrial practice, namely correcting deviations from design specifications during manufacturing so that parts and assemblies are produced on target tolerances specified.

Providing feedback from inspection (verification) to the manufacturing process is demonstrated to be feasible (Hoffman, 1982, Wirtz, 1988, 1991, among others). This requires a fresh start for the definition of the tolerance specification itself because design specifications must be unambiguously convertible to tooling and setup specifications, inspection specifications, and process corrections. As pointed out earlier, the current standards have limitations. Vector tolerancing scheme proposed by Wirtz illustrates how this can be accomplished through vectorial representation of the position, direction and size along with the corresponding tolerance zones. This scheme leads to matrices in a coordinate system for nominal dimensions and tolerances. During inspection, a matrix of the real surface is obtained from inspection data on the part in terms of the deviation matrices (for mean value). The difference between the real surface matrices and the specification matrices provides the correction matrices to bring the manufacturing process into control through appropriate adjustment of the tool's position, direction, and size. The scatter matrix obtained from the standard deviation data provides information for technological control of the process in reducing the variability of the process. A similar but somewhat limited scheme involving only +/- tolerances has been recently demonstrated for automatically converting design specifications to manufacturing setup and NC cutter path generation (Anderson et al., 1991). These approaches indicate a promising direction for further development because of the type of practical application mentioned above. However, many of the limitations stated above must be overcome.

TRADITIONAL INDUSTRIAL PRACTICES FOR SPECIFICATION, REALIZATION AND VERIFICATION (INSPECTION) OF TOLERANCES AND DEVIATIONS

The roots of traditional practices can be traced back to an era when increasing specialization produced distinct separation between design, manufacturing planning, production, and inspection functions. Although the specification of tolerances and deviation was recognized to be a crucial design function requiring engineering experience of the entire product life cycle, the actual tolerancing and specification of deviations of the drawing was relegated to drafting. The design engineers who conducted design analysis using increasingly sophisticated analysis methods rarely took interest until there was a field-failure to warrant so. The traditional engineering cycle is based on the design-build-field/test-modify iteration. Indeed, the very language of conveying designs which requires preparation and interpretation of drawing, tolerancing and deviation conventions have been steadily declining from curricula of most U. S. engineering schools for the last three decades. The task of drafting and documenting the drawings of the designs prepared by engineers has been relegated to a cadre of drafts-persons (so-called designers) with a 2-year junior college program. This tradition continues with the advent of CAD systems which still are essentially computer-aided drafting systems.

The specific value of tolerance and deviation in the traditional design is settled by a compromise between the designer's desire to tighten the tolerances because of fear that manufacturing will relax them anyhow and the desire of manufacturing to relax the tolerances to minimize the tightening of controls over an available manufacturing process. Inspection is traditionally a 'watchdog' organization policing quality through verification of actual production to weed-out parts and assemblies that do not satisfy the specified tolerances and specifications.

Although the limitations of the traditional practices are well known, it is important to briefly review them because they contrast against the recent advances as demonstrated by the case studies presented later.

Statistical Quality and Process Control

The task of inspection involves verification of manufactured parts and assemblies to the conformance of design specifications of shape, size, form, finish and other dimensional and non-dimensional tolerances per standards. Nonconforming parts/assemblies either must be reworked or rejected. The manufacturing processes producing nonconforming parts and assemblies must be reset/adjusted to bring them back into producing conforming parts/assemblies. The methodologies of statistical quality and process control have been developed and successfully applied to inspection processes since the 1930's (Shewhart, 1931, Evans, 1974,1975). It is important to review how capabilities and limitations of these methods have brought about the recent advances in specification and control of tolerance and deviations.

Variabilities of Manufacturing Processes

The current tolerancing standards recognize the fact that the inherent variabilities in materials and manufacturing processes produce part shapes and dimensions that deviate from the ideal geometric specifications and hence provide a tolerance zone as specified around the nominal dimension. However, these standards do not include any statistical basis for tolerancing. The traditional methods for tolerance analysis (so-called tolerance charting schemes for computation of assembly gap tolerance) and synthesis (allocation of tolerances from assembly to individual parts) consider only discrete tolerance evaluation of one-dimensional linear or nonlinear assemblies. These methods involve "worst-case" scenarios which invariably lead to tighter than necessary part tolerances and more expensive manufacturing. Statistical methods for tolerance analysis and synthesis are closer to reality provided the distributions are known (eg. Bjorke, 1973; Chase et al., 1990, among others). Despite the availability of some of these methods, few have been implemented because reliable data on distributions have been rarely available during design stages.

Control Charts and Other Statistical Methods

The methodologies for statistical control of quality and process have evolved to maintain the parts produced within the tolerance specifications. It is well-recognized that for the parts and assemblies to satisfy the specified tolerances, the manufacturing processes must be selected, set-up, and maintained well within the tolerable limits in view of the tendencies of most manufacturing processes to drift and exhibit dispersion due to the inherent variabilities of workpiece and tool materials, and processing parameters. Each manufacturing process has a characteristic variability that leads to the specific types of distribution of the realized part or assembly dimensions, deviations of form, and surface finishes. A systematic methodology for inspection, including sampling plans, and determination of the parameters of distribution such as estimates of mean, variance and skewness has been the concern of statistical inspection techniques. One of the powerful tools of statistical quality control is *control charts* in which critical parameters of the process are tracked through appropriate sampling and plotted for revealing whether or not the process parameter is within the specified tolerance limits. $\overline{X}$ (average) and R (range) charts are among the most commonly used; p (fraction defective), np (number defective), u (average number of defective per unit of sample) and other specialized charts have been introduced (Small/AT&T Handbook, 1956). Although the control charts have been the main tool of statistical quality control, design of experiments (Fisher, 1935), time series analysis and other statistical testing methodologies have been in use, initially for chemical processes and more recently, for mechanical products and processes. Specialized services of statistician are generally required to assist engineers conduct industrial applications of these methods.

Control Charts and other statistical methods became the domain of inspection function which has not been a separate academic discipline. Until recently, the status of quality control and inspection functions was low with respect to that of production and design. Producing higher quality parts was considered to require costly materials and processes, and expensive inspection, rework and rejects. The time spent in inspection was considered to be a loss to productivity. The traditional cost accounting systems did not keep track of the cost of rejects and rework or the cost of inspection; it was absorbed in the plant overhead. These and other limitations of the traditional practices were revealed only when U. S. Corporations faced tough competition from imported higher quality competitive products, primarily from Japan.

A COMMENTARY ON THE CASE STUDIES

One of the key reasons for the high quality of Japanese products is the adoption of statistical quality control methods by its industry. The story of how Japanese industry passionately adopted the teachings of Deming, Juran and other U. S. quality control consultants is well known by now. Other well known Japanese innovations are Just-In-Time/Kanban, Quality Circles, the philosophy of Total Quality Control, and Kaizen (an umbrella for all concepts and tools for continuous improvement), among others. Before commenting on the case studies, it is necessary to briefly review the more recent innovations of Robust Design, Quality Loss Function, and Quality Function Deployment because the leading U. S. Corporations involved in these case studies have successfully adopted and improvised on these methodologies (Ealy, 1988). During the past few years, most engineers have had the opportunity to be exposed to these methodologies through professional and popular media, and hence I shall confine my remarks only to the points pertinent to the case studies.

Quality by Design

Utilizing the techniques imported from the U. S. and building on additional tools, the methods of quality control evolved into quality engineering in Japan. Soon after the adoption of control charts for product and process control, the Japanese mechanical, electro-mechanical, optical products, and other industries began to tighten the requirements from defective parts per thousands to defective parts per millions. This essentially meant dropping the traditional Three Sigma limits and Accepted Quality Levels (AQL) followed by traditional quality control practices developed in U. S. A. during the 1930's and 1940's, and adopting higher Sigma limits for components that make the product. Tightening these limits means using more expensive materials and processes.

Important as it is, the statistical quality control as implemented through control charts and other statistical tools, does not inherently improve quality; it allows weeding out of defective parts/ assemblies prior to shipping the product to the customer. These techniques can identify the root causes for deviations which when corrected would reduce the deviations. Eliminating the root causes for variation also incurs additional cost.

Some of the leading Japanese quality researchers and consultants, Ishikawa, Taguchi, and Akao among others, realized that *meeting specifications not only does not necessarily guarantee quality; it also does not automatically guarantee customer satisfaction with your product.* The product specifications, including tolerances, material specifications, and process specifications and deviation limits are not the true quality characteristics (Ishikawa, 1972, 1977). They are substitute characteristics whose relationship to the true quality characteristics must be determined. Just meeting specifications of tolerance and deviations, no matter how well they are based on experience, does not, therefore, guarantee customer satisfaction.

Quality Loss Function and Robust Design

Taguchi observed that just conformance to specification does not produce the same degree of satisfaction; the closer the product is to the target within the tolerance band the higher the satisfaction. Taguchi, however, avoided getting into the subjective 'value' of quality to the customer; he did not consider this to be an engineering problem. According to him, quality product performs its intended functions without variability, and causes as little a loss as possible through harmful side effects, including the cost of using it. He pointed out that the true cost of quality is the loss to the customer and society at large due to the poor performance of the product (Taguchi, 1986, 1987; Phadke, 1989). The real cause of this loss is the variability of product performance as the product deviates from its target even though it may conform to the tolerance and deviation specifications. *"When a product fails, you must replace it or fix it. In either case, you must track it, transport it, and apologize for it. The losses are much greater than the cost of manufacturing it; none will recover your reputation leading to the loss of market share..." (Taguchi* and Clausing, 1990). Taguchi introduced a quadratic loss functions known as Quality Loss Function in which as the product characteristics depart from the target, the quality loss to the customer grows quadratically with the deviation. A product that produces the least amount of loss is said to be Robust. During the introduction of these methodologies, Taguchi asked an important question: *Can products be made less sensitive to the variations in the user environment as well as to the variations in manufacturing process by design?*

Taguchi introduced a systematic methodology for design of Robust products in response to this question. It consists of (1) *Parameter Design:* setting the values to the parameters of the product (System, subsystem, components or parts) through minimization of S/N (signal to Noise) ratio; (2) *Tolerance Design:* setting tolerances to minimize the loss caused by not meeting the target within the specification; and (3) *On-line Quality Control:* involves process diagnosis and adjustment, prediction and correction; measurement and corrective actions, however, can only deal with the unit to unit variations.

Parameter design, involves a series of tests based on an orthogonal array in which the ratio of the signal of the main function of the system to the noise is determined at specific sets of combinations of the levels of control and noise variables. The noise variables are those of the manufacturing process, user environment as well as those from degradation of the product due to its use. In parameter design the objective is to start with inexpensive materials and components which may exhibit large variability and determine those combinations which are least sensitive to the variations in use or in manufacture. In tolerance design, the expense of reducing variability of critical components is weighed against the additional cost through the quality loss function. Taguchi parameter design is applicable to dynamic characteristics as well.

Taguchi points out that the losses due to variation in the user environment can only be minimized by design. During design stage, the impact of losses due to material, product degradation, and manufacturing process can also be minimized. However, once the design is set, robust design methods during manufacturing engineering and production can only be applied to minimize the impact of unit to unit variation caused by material and process variables.

Taguchi methods have come under criticism from the statisticians who point out that the use of orthogonal design ignores interactions and that the S/N ratio is not necessarily an appropriate objective function in all cases (Box, Bisgaard and Fung, 1989). This criticism may be valid; after all, the test designs used in the Taguchi method are essentially those from the well known designs of factorial and fractional factorial methods of Design of Experiments methodology introduced by Fisher and developed by many others. However, it is the opinion of many design engineers I encountered during these case studies that the Quality Loss Function and Robust design methods represent a more viable design methodology than the traditional design practices. The value of the parameter and tolerance design methods is in constantly simplifying and improving the subsystems and components through a series of tests so that only a single prototype is necessary. By testing the physical subsystems and components of the design, the guesswork in design is eliminated. Tolerances and deviations are not set just on past experiences or guesses but on the cost trade-offs with the loss function.

Quality Function Deployment

While the methods of Robust design and Quality Loss Function are capable of evaluating and minimizing the deviation of the design parameters from the targets, the targets themselves must be obtained from the customer requirements. A methodology for determining what characteristics satisfy customer was evolved out of the work on quality tables at Kobe Shipyards, Mitsubishi Heavy

Industries Company, Japan in 1972 (Kamizawa, Ishizuka and Akao, 1978). The quality tables were used to bring marketing and design functions to simultaneously focus on determining customers' requirements, including functions, features, and quality issues. The concept of quality table evolved into Quality Evolution System (QES) in which design quality for finished products is established by the interpretation of the user's demands into substitute quality characteristics for functional components and individual production process elements (Mizuno and Akao, 1978). The authors credit Juran and other work on Quality Assurance, Value Analysis, Failure Mode Analysis, and other methodologies for providing a starting point for the concepts of QES. The methodology for taking customers/users requirements and competitive benchmarking into account for setting specific functional and esthetic design targets came to be known as Quality Function Deployment (QFD) (Akao, Y. editor, 1987). The concepts of QFD (so-called House of Quality) are explained through an example of door closing effort at Toyota Motor Company (Hauser and Clausing, 1988). In QFD, the product attributes which the customer wants are ranked by an inter-functional team of individuals from marketing, design, manufacturing, and strategic planning, and ranked in order of preference and weighed against those of the competitors through benchmarking. Specific competitive functional targets and tolerances are set for each important product characteristics. The product targets are resolved into part targets, process planning targets, and production control targets by following similar houses at each resolution step. This method has become well accepted at a number of U. S. Corporations. Because it focuses on specific functional targets for design of competitive products, QFD is capable of providing viable inputs to Robust parameter and tolerance design of products.

A Commentary on Case Studies

The criteria for the selection of case studies I adopted was based on how extensive is the adoption of contemporary methodologies for setting specific tolerance and deviation targets to ensure 'quality by design'. The award winners of Malcom Baldridge National Quality Award: "Quest for Excellence" were considered to be the potential candidates because of the rigorous scrutiny these companies must undergo according to the award criteria. First, the available case studies on the contemporary methodologies (ASI Taguchi symposiums volumes, GOAL/QPC Seminar proceedings, Taguchi Methods Applications in World Industry, edited by Bendell, Disney and Pridmore, 1989, as well as the company in-house cases) were reviewed to determine what type of case studies should be included in the investigation. This was a difficult choice to make because there are varieties of applications and interesting cases available from many sources. I have chosen the case studies based on my judgement about what should be brought to the attention of the active researchers in mechanical tolerancing, the ASME codes and standards community, and NSF project managers. I have assumed little or no background in contemporary methodologies on the part of the readers in the belief that the readers already familiar with the topic will forgive the repetition in the interest of reviewing some of the potential research and implementation challenges. The comments below on each of the case studies are intended to show the relevance to the topic.

Producibility, Process Capability and Six Sigma Quality

Producibility, in the simplest sense is ease of manufacture. Motorola defines it as *"the ability to reproduce, identically and without waste, units of product so that they satisfy customer physical and functional requirements (quality, reliability, performance, availability, and price) and also satisfy Motorola business goals"* (Harry and Lawson, 1988). Based on the numerous application cases over the last half decade, a methodology for assessing producibility of a design has been evolved at Motorola. As pointed out by Mikel Harry in the case study (Harry, in this report), producibility is measured quantitatively during the product and process design stages so that different design alternatives can be compared. Producibility and quality are complementary concepts; process capability using Six Sigma limits instead of the traditional Three Sigma limits has been well established at Motorola.

Process capability is determined from control charts data which can reveal the underlying characteristics of the distribution of the variability as well as the drift of the process over time. Process capability index which indicates how capable is the process of producing parts within the specified tolerance has been introduced and is widely used in several industries. At Motorola, Cp = (USL-LSL)/six sigma; where, USL and LSL are upper and lower engineering specification limits (i.e. tolerances); Cpk = Cp(1-k), where k is a measure of the process drift with time, it amounts to be 0.5 when the process drift is 1.5 Sigma. For Six Sigma quality, Cp = 2 and Cpk = 1.5; Six Sigma quality gives 3.4 ppm versus 66,803 ppm opportunities for nonconformance for Three Sigma traditional limits.

Since a product consists of a number of components and a number of manufacturing and assembly setups, producibility is computed from the probability of simultaneously satisfying all the product specifications. It is computed by taking into account the independence and mutual exclusivity of the events that lead to nonconformance. The application of this analysis requires reliable data on process capability of individual components and processing steps; at Motorola, a great emphasis is placed on establishing process capability data from vendors/subcontractors and in-house production facilities for use during product design stages. Optimization of product and process parameters is done using classical design of experiments (eg. Harry, 1989). Time series analysis and application of Chaos theory are being explored as presented by Mikel Harry in a panel session at the ASME CIE Conference, 1989 (also, Harry and Prins, 1991).

Application of QFD to Engineering Specifications and Product Development

Ford engineers applied QFD to answer a crucial question: *How can the specifications of a component (in this case, a Variable Assist Power Steering valve) be written in a manner that assures customer satisfaction with the steering performance?* Upon setting up the QFD (House of Quality) table, Ford engineers compared a series of Engineering Specification tests with the customer requirements and found that some of the important customer requirements were not addressed by the existing Engineering Specification tests (Smith, in this report). This led to additional Engineering Specification tests such as thermal cycling and impulse to meet specific quantitative targets and tolerances for the major customer requirements. Furthermore, specifications for supplier and sub-supplier of the parts and components of the valve followed from the QFD of the valve.

At Ford, QFD process has been adapted into a systematic methodology as an integral part of the product development process as demonstrated by the CIII Power Steering Pump example.

A massive training of Ford in-house and supplier personnel in methodologies of quality control is an on-going activity at Ford. Extensive use of QFD and Robust design methodologies has been credited with a major portion of quality improvements at Ford and its suppliers.

Application of Robust Design and Quality Loss Function

A comprehensive example of how parameters of the paper feeder mechanism were obtained by applying Robust design and Quality Loss Function methodologies introduced by Taguchi is presented by Xerox (LaValee, in this report). This case study shows how a significant advantage can be gained by adopting robust design methodologies (Taguchi methodologies) instead of the traditional design methodologies. By focusing on the main function of the mechanism and how it should perform, the range (window) was significantly broadened over which *misfeed* and *multifeed* of papers will not occur despite the variabilities during usage and during the manufacturing process. The use of S/N ratio and the orthogonal array of experiments is well illustrated in this case study.

Another example of the application of these methodologies was chosen from extensive series of applications at AT&T over the last decade. Just as AT&T was one of the pioneers in the application of statistical quality control methods (AT&T, 1956), it has been one of the earliest adopters of Taguchi robust design methodologies. Of the several examples, the application of the methodology to the improvement of the performance of computer hardware/UNIX operating system software was chosen as the case study for illustrating how these methodologies have been applied to such systems (Phadke, in this report). It is interesting to note that the experiments in this case were conducted without any down time on the system; UNIX utilities were used to keep track of the data while the system was in operation. Adoption of such methodologies may significantly improve the performance of CAD/CAE/CAM computer systems used by design and manufacturing engineers.

By now the extensive applications of the robust design methodologies has spread to several U. S. companies as evidenced from the case studies presented at the yearly ASI symposium on Taguchi methods. These methods are becoming a tool for design and manufacturing engineers for improving the prototype designs of products and processes.

Organizational Change for Competitive Response

How did such a significant change from traditional to robust design methodologies take place in these corporations? The answer to this question can be found in the way Xerox company responded to the competitive challenge from Japanese copier manufacturers (Bebb, in this report). Upon reading the popular account published in the book *Xerox, American Samurai, by Gary Jacobson and John Hillkirk, McMillan, 1986*, I decided to investigate how the robust design methodologies were adopted in Xerox and what organizational changes were made to regain the competitive edge, particularly the way tolerances and deviations were specified and controlled. The concept of how design latitudes and manufacturing variances can be determined and managed throughout the design-prototype cycle are well illustrated in the case study. This case study focuses on the important issues of how the design process was changed. I can report that based on my interviews with the practicing design engineering teams, the methodology of robust design through testing prototypes of systems and subsystems using Taguchi methods is well established at Xerox. The principal organizational change was that the chief designer of each product was made responsible for carrying the product development through the production of the product instead of handing over to manufacturing after design as in the past. The task of introducing the Taguchi methodologies was carried out by a support group called Design Institute within Xerox.

This case study also points out how the subject of quality engineering has been neglected in U. S. corporations as well as in engineering and business schools. Xerox has had a valid benchmark to compare its organization and practices with those of its Japanese sister company, Fuji-Xerox. While working on designs side by side, Fuji-Xerox design engineers were found to constantly keep on tuning the design-prototype through a series of robust design tests; the Xerox design engineers, on the other hand, tended to run only a few tests on completed prototype designs. The Fuji-Xerox design engineers were using the robust design tests as a tool for improving the design; the Xerox design engineers tended to use it as a verification method for their design. In a recent symposium,

I had the opportunity to question some of the corporate officers at Fuji-Xerox on the difference between the two organizations. The principal difference was that the Fuji-Xerox corporate managers had risen to their position through the engineering ranks. They themselves had first hand experience as product mangers. Another principal difference between the two organizations, according to one Fuji-Xerox corporate manager I interviewed, was *the way failures are perceived and treated.* In Fuji-Xerox, recalling from his personal experience, he pointed out that he was allowed to fail and given a chance to learn from his mistakes (Tsuchiya, 1991). Because of this tendency, he found that the Fuji-Xerox product managers were more experienced than their counterparts at Xerox. Another difference was that while Fuji-Xerox engineers tended to uncover problems during design reviews, the Xerox engineers tended to suppress problems during design reviews. He also found that the senior management at Fuji-Xerox had understood and taken the task of implementation of Taguchi robust design methodologies seriously. They considered uncovering problems to be the storehouse for future improvements, and hence, to be encouraged. He further cautioned that once launched, a product development process should not be interrupted just because a competitive product has been introduced. Also, only well established technologies should be used for product development. The competitive edge is in the way product development process is managed; engineers at both organizations were equally capable.

Towards the Development of a Comprehensive Methodology for Tolerance Analysis and Synthesis

The fact that the traditional tolerance analysis and synthesis practices in industry still treat the problem by worst case analysis and that the methods of statistical tolerance analysis and synthesis are rarely applied have been pointed out earlier in this paper. The probabilistic approach to engineering designs has been recognized as a viable method where uncertainties in the system, environment, and manufacturing processes can be taken into account. The case study from Xerox is presented as one attempt toward this goal (Parks, in this report). This case study present a framework in linking the customer dissatisfiers to the product design through flow of variances from the customer requirements to individual part tolerances and deviations, including considerations of the Taguchi Quality Loss Function. As the author of this case study points out, the key to implementation of such a methodology is the predetermination of the variances and probability distributions of the individual phenomena introducing the variance in the main function of the design. If successful, such methods will allow tolerance analysis and synthesis to be modeled and investigated prior to making physical prototypes. With the rich examples of actual product design and implementation cases, Xerox researchers stand a better chance in determining this feasibility than others without such access.

SOME OBSERVATIONS ON RESEARCH NEEDS AND IMPLEMENTATION CHALLENGES

A review of these case studies provides a unique opportunity to researchers, standards communities, and practitioners to consider what the case studies mean to the tasks of (a) building sounder theories for tolerances and deviations, (b) eliminating ambiguities from current standards, and (c) implementing successful competitive practices for improving functional quality of products. Some who have been at the leading edge may find these case studies a reinforcement to their own experience. Others who have been considering narrowly focused issues of tolerancing may find these case studies out of the realm of what they thought tolerancing should entail. One thing for sure, these case studies are an eye opener. They demonstrate that in implementation all relevant issues must be faced with whatever tools are available. As these case studies demonstrate, the leading corporation learned to adopt new tools by being vigilant about what their competitors were doing. To them the name of the game is to design and manufacture high quality competitive products. Whether or not theory or standards are adequate is a mute point. In the interest of promoting thoughtful discussions, I have offered my observations about the case studies through pros and cons of the following questions:

(1) How should the successful methodologies from these case studies be further improved to make them even more effective for implementation?

This is perhaps the most relevant question for practitioners in industry. It is also relevant to research and standards communities because their results must be ultimately implemented in industry. The more relevant and timely the response from these communities to the challenges faced by industry, the more competitive the industry is likely to be.

The methodologies of QFD, including the definition of customer's functional requirements, competitive benchmarking, reverse engineering, setting of competitive targets for quality, cost and rapid new product introduction need to be put on a firmer foundation so as to increase the ease and speed of their application. These methodologies are a means to bringing otherwise separate functional viewpoints of business strategy, marketing, engineering, manufacturing, production, and service. These methodologies should provide valuable inputs to concurrent engineering and rapid prototyping.

The methodologies of Robust Design and Quality Loss Function have already proven to be effective in industry, as these and many other case studies demonstrate. Application of these methodologies has shown that the design process itself must be improved. The outmoded traditional design process cannot match the results and speed of the Robust design process. Compared to the traditional design process in which delays and cost overruns due to several prototypes and engineering changes are a common place, the Robust design process requires only a single robust prototype, greatly accelerating the introduction of a quality product into production without the need for many engineering changes. It is clear that these improved methodologies are crucial to the success of concurrent engineering.

There has been some attention paid to establishing a better statistical foundation for these methodologies; this is a welcome trend. However, the real benefit of these methodologies will come from integrating them in the design tools that are easily accessible to concurrent engineering teams.

(2) What should be done to overcome the implementation barriers posed by traditional corporate organizations?

As seen from the case studies, there are three kinds of implementation barriers that must be overcome:

(a) Education and training of methodologies: The case studies show that anything short of massive and sustained on-site educational and training programs are likely to be ineffective. Motorola has established Motorola U. which provides at least 5 weeks of training to each employee besides conducting numerous on-going courses and workshops on-site (Wiggenhorn, 1990). Ford has focused on supplier education through ASI (ASI scheduled training seminars). Xerox has had design institute to educate and train design engineers. AT&T Bell labs has had a long tradition for

in-house training and consulting organization for its operating units. There are other prominent examples in U. S. industry as well. The point is education and training is an on-going and a massive task to which these leading organizations are committed. However, compared to the extensive training in quality control methods which Japanese workers go through, these efforts are still limited (Box et al., 1988). What should be the course work and how the subjects should be introduced is worth examining to make these in-house efforts more effective. Instead of focusing on individual training alone, more group training and team building workshops should be developed to carry what is learned into thework place.

(b) Organizational changes in the design process: The traditional functional barriers between design, manufacturing, and production were overcome by integrating the task of new product design, prototyping, and manufacturing engineering under a single product design manager as demonstrated at Xerox. Unless it is appreciated that the Robust design methodologies cannot be implemented by simply adding more computer aids to the existing traditional organization, the organizational change will not stand a chance. Change happened only after a crisis which put Xerox copier business at risk of losing to the competitors. Could the change revert back to the established traditional functional organization once the crisis is over? Yes, it could, because the traditions have many vested interests and because any change is a threat to those committed to the status quo. Introduction of concurrent engineering is a good reason to nucleate small multi-functional teams to carry the product from conception to the start of production. A few leading U. S. corporations have made significant progress towards this process.

(c) Commitment for sustained change from enlightened top management: This is perhaps the most important barrier to overcome. A change may originate at any level within a traditional organization; however, without the sustained commitment from the top management, the above two barriers cannot be overcome. To obtain this top management commitment, the top management first must take time to understand the improved methodologies, and then must have a strategy for its adoption with appropriate resources assigned to it. To realize how they must change, the U. S. corporate top management has only to benchmark themselves against their counterparts in Japan, as Xerox did in the 1980's (Jacobson and Hillkirk, 1986).

(3) How should the fresh engineering graduates be better prepared for industry?

U. S. engineering education has steadily moved away from empirical engineering practices to science-based narrowly specialized curricula (many subdisciplines in each branches of engineering education) during the last three decades. Fresh engineering graduates, especially those entering into industry after post graduate education have had to undergo a cultural shock in which they are thrusted into the real world of engineering empiricism and practice for which they had little preparation (Fry, 1991); 'co-oping' or doing summer jobs in industry helps but does not by itself assure exposure to industrial practices. In narrowly focused specialities, such a transition has not been drastic. However, in the area of design, the fresh graduates faced a most difficult time in relating their meager experience at school to the broad and far reaching implications of design decisions and design practices in industry. Realizing this must be changed, several engineering schools are now reexamining their design curricula (Dixon, 1991). It is high time that this question be answered, not by trying to juggle current engineering curricula, but by taking a fresh view of what should be done. Industry, engineering schools, and business schools must work together to design such curricula (as indeed is already happening among others, at M. I. T. Leaders for Industry program). However, there is a lot to be done to bring the question into close scrutiny. To answer this question, we must reexamine the changes that must occur in the current hiring, promotion, tenure, teaching and research responsibilities, and most importantly, industrial experience of the professors. NSF should actively promote discussions on this question through industry-university workshops that develop alternatives and that promote a meaningful discussion of the topic.

(4) What should be the goals, scope and objectives of tolerancing theories?

Since the research needs for theories of tolerancing have been well stated in the report of the last international workshop on Mechanical Tolerancing (Tipnis, et al., 1989), I shall not repeat them here.

As I have pointed out earlier in this report, a comparative analysis of the current mechanical tolerancing theories is not yet available; this must be done to know what the different approaches are. Most well known current approaches are directed towards purely geometric aspects of representation of solid rigid objects. This is challenging enough by itself, according to those involved in the development of the theories. However, it falls short of addressing the real physical objects that a designer must deal with. The relationship between tolerances and design functions should not be neglected. Also, verification (inspection) metrological community has realized that purely geometric reasoning cannot address the real physical aspects of the verification issues.

Taguchi's Robust design methods have shown a rational approach to determining the tolerances and design parameters through Quality Loss Function and through robust design tests on real prototypes for maximizing Signal to Noise ratio. Also, there are attempts to establish propagation of tolerances and deviations from customer to the individual parts in the product assembly as has been reported in one case study (Park, in this report). Such probabilistic approaches coupled with the computational mechanics models may lead to a practical methodology. Specifications of tolerances should be verifiable through geometry/material computer simulation of physical aspects prior to making a real prototype.

The concepts of Features-based design are currently actively pursued in the CAD community. However, these are mostly based on geometry alone. The idea that purely geometrically based definitions of feature will not be sufficient has now been appreciated. The concepts of functional features and manufacturing features and their connection to tolerances have been proposed (Tipnis, 1991). Such concepts may lead to representations of functional requirements that are an integral part of the product design model, thus eliminating the design notes and labels which are subject to misinterpretations.

(5) Should the standards activity be limited to purely geometric tolerancing of rigid mechanical objects as in the past or should a new agenda be charted?

Work is already in progress for putting current geometric tolerancing on a firmer mathematical footing as is evident from the ANSI Y14.5.1 subcommittee efforts mentioned earlier in this report. Cooperation with ISO and other national and industry standards communities is sought in this effort. Also, the efforts towards improved algorithms which may serve as standards for CMM are underway at different ANSI and ISO standards committees. The mechanical tolerancing standards community is closely watching the developments of STEP/PDES, CALS and ISO 9000 because of the potential impact of these development on the specification of mechanical designs. These no doubt deserve the most attention from the mechanical standards community. However, one of the real gaps pointed out in the last international workshop on mechanical tolerancing (Tipnis, et al., 1989) still remain unaddressed: namely, how to specify tolerances on seals, O rings, gaskets, lubricants/sealants, plastics and fiber materials/limply materials. Non-rigid solids are often the most crucial weak link in the mechanical design as we have painfully learned from the Challenger Shuttle disaster. Elastic, plastic, and viscous deformations and shape changes that dramatically alter the interfaces between mechanical components must be brought under the realm of mechanical tolerancing standards. Also, recognizing that the real physical solid objects exhibit variance in shape, size, form and surface characteristics, and are subject to elastic, plastic, and viscous deformations, the scope of the mechanical standards should be broadened to include such aspects. ASME and NSF could initiate exploration of these possibilities through a series of workshops or dialogues on the topic.

In closing, I welcome readers of this report to form their own observations about the case studies from their own unique perspectivs and view points and comment on the questions raised. I would welcome any comments forwarded to me.

REFERENCES

Akao, Y., (1987), *"Quality Deployment"*, edited articles, G.O.A.L./Lawrence, 1987.

Akao, Y., (1987), "Quality Deployment," A series of articles, Growth Opportunities Alliance of Lawrence, Inc., Lawrence, MA., 1987, pp. 1-1 to 12-12.

Akao, Y. Ishizuka, I., and Kamizawa, N., (1978), "Quality Evolution System and FMEA," Proceedings of the International Conference on Quality Control, 1978, Tokyo, pp. B4.19-B4.28.

Anderson, D., Tang, C. T., Kak, A., and Mitchell, O.R., (1991), "A. Quick Turnaround Cell", Presentation at the Feature-Based Design and Manufacturing Workshop, NCMS sponsored, Gatlinburg, TN, June 19-20.

ANSI Standard Y14.5M - 1982, *"Dimensioning and Tolerancing"*, U. S. Standards Institute, NY, (1982).

ANSI Standard B89.1.12M - 1985, *"Methods for Performance Evaluation of Coordinate Measuring Machines"*, United States Of America Standards Institute, NY, (1985).

Bjorke, O. 1978. *"Computer-Aided Tolerancing"*. Trondheim, Norway: Tapir.

Bendell, A., Disney, J. J., and Pridmore, W. A., (1989)" Taguchi Methods: Applications in World Industry', IFS and Springer Verlag, 1989

Box, G. E. P., Bisgaard, S., and Fung, C. (1989) "An Explanation and Critique of Taguchi's Contribution to Quality Engineering" in " Taguchi Methods: Applications in World Industry', IFS and Springer Verlag, 1989, Bendell et al. editors, pp. 359-383.

Box, G. E. P., , Kacker, R., Nair, V., Phadke, M., Shoemaker, A., and Wu, C. F. J. (1988) " On Quality Practices in Japan", Quality Progress, 1990

Chase, K. W. et al. (1990), " Recent Developments in Tolerance Analysis Software for Mechanical Assemblies, Part A: Two Dimensional Modeling", Dept. of Mech. Engg., BYU, Provo, Utah.

Dixon, J. R., (1991), " A Review and Critique of the NRC draft Report entitled Improving Engineering Design", Proceedings of Design Productivity International Conference, Ragsdell, K. E. and Holt, T. editors, Feb. 3-9, Hawaii, 1991, vol. 2, pp. 967-971.

Ealey, L. A., (1988), *"Quality by Design: Taguchi Methods and U. S. Industry"*, ASI Press, 1988.

Evans, David H., (1974). "Statistical Tolerancing: The State of the Art, Part I: Background," Journal

of Quality and Technology, 6(4), pp. 188-195.
Evans, David H., (1975). "Statistical Tolerancing: The State of the Art, Part II: Methods for Estimating Moments," Journal of Quality and Technology, 7(1), pp. 1-12.
Evans, David H., (1975). "Statistical Tolerancing: The State of the Art, Part III: Shifts and Drifts," Journal of Quality and Technology, 7(2), pp. 72-76.
Feng, S. C. and Hopp, T. H., " A Review of Current Geometric Tolerancing Theories and Inspection Data Analysis Algorithms', NIST Report # NISTIR 4509, February 1991.
Fry, Don, (1991), "Learning The Ropes: My Life As A Product Champion", HBR, Sept.-Oct, 1991, pp. 46-57.
Fisher, R. A. (1935), *"The Design of Experiments"*, Oliver and Boyd, Edinburgh, 1935.
Harry, M. J. (1986). "The Nature of Six Sigma Quality". Motorola University Press, Motorola Inc., Schaumburg, Illinois.
Harry, M. J. and Lawson, R. J. (1988). "Six Sigma Producibility Analysis and Process Characterization." Publication Number Six Sigma-3-03/88. Motorola University Press, Motorola Inc., Schaumburg Illinois.
Harry, M. J. and Stewart, R. (1988). "Six Sigma Mechanical Design Tolerancing." Publication Number Six Sigma-2-10/88. Motorola University Press, Motorola Inc., Schaumburg, Illinois.
Harry, M. J. (1989), "PCB Plated Through Hole Optimization: A Case Study in SPC", EIPC Conference, 1989.
Harry, M. J. and Prins, J. (1991). "The Vision of Six Sigma: Mathematical Constructs Related to Process Centering." Publication Pending. Motorola University Press, Motorola Inc., Schaumburg, Illinois.
Hauser, J. R. and Clausing, D. (1988), "The House of Quality", HBR, May-June, 1988, pp. 63-73.
Henzhold, G. (1991), 'The Principle of Independency and Geometric Tolerances', Proceedings of CIRP Intl. Working Seminar on Computer-Aided Tolerancing, Ham, I and Lehtihet, E. A., editors, Penn State, May 16-17, 1991.
Ishikawa, K., (1977), "Quality Analysis", 1977 ASQC Technical Conference Transactions, Philadelphia, pp. 423-429.
Jacobson, Gary, Hillkirk, John, (1986), *" Xerox, American Samurai"*, McMillan, 1986.
Juran, J. M., Gryna, F. M., and Bingham, R. S. (1979). Quality Control Handbook. New York, NY: McGraw-Hill Book Co.
Kamizawa, N., Ishizuka, I., and Akao, Y., (1978), "Quality Evolution and FEMA", ICQC, Japan 1978, Proceedings, pp. TS. B4-19 to B4-28.
Mizuno, S. and Akao, Y, (1978), "The Evolution of Quality Function-An Approach to Company Wide Quality Control", JUSE, 1978.
Phadke, M. S. (1989), *"Quality Engineering using Robust Design"*, Prentice Hall, 1989
Requicha, A. A. A. (1983), "Towards a Theory of Geometric Tolerancing", International Journal of Robotics Research, 2(4):45-60, 1983.
Requicha, A. A. A. (1984), "Representation of Tolerances in Solid Modeling: Issues and Alternative Approaches", Solid Modeling by Computers, Pickett, M. S. and Boyce, J. W. editors, Plenum Publ., 1984, pp 3-22.
Shewhart, W.A. (1931). *"Economic Control of Quality of Manufactured Products"*. D. Van Nostrand Company, Inc.
Small, B. B. et. al., *"AT&T Statistical Quality Control Handbook"*, 1956.

Srinivasan, Vijay, (1991), "A Geometer Grapples with Tolerancing", Proceedings of CIRP Intl. Working Seminar on Computer-Aided Tolerancing, Ham, I and Lehtihet, E. A., editors, Penn State, May 16-17, 1991.

Taguchi, G., *"Introduction to Quality Engineering"*, Asian Productivity Press, 1986

Taguchi, G., Elsayed, E. A., and Hsiang, T., *"Quality Engineering in Production Systems"*, McGraw-Hill, 1989.

Taguchi, G. and Clausing, D. (1990), "Robust Quality-Taguchi Methods", HBR, Jan/Feb., 1990, pp. 65-75.

Tipnis, V. A. et. al., (1989) "Research Needs and Technological Challenges in Mechanical Tolerances", Results of an International ASME/NSF Workshop, October, 1988, Orlando, FL, ASME Report CRTD-15, 1989.

Tipnis, V. A., (1991) , "Research Needs of Feature Based Technology: The Concept of Functional and Manufacturing Features", CIRP, Part II, August, 1991, Palo Alto.

Tsuchiya, Motohiko (1991), "Efficient Development of New Products which Achieve Customer Satisfaction", Proceedings of Design Productivity International Conference, Ragsdell, K. E. and Holt, T. editors, Feb. 3-9, Hawaii, 1991, vol. 1, pp. 27-31.

Wiggenhorn, W., (1990) " Motorola U.: When Training Becomes an Education", HBR, 1990, Number 4, pp. 71-83.

Voelker, H. B., (1988), "Modeling the Design Process", in Design and Analysis of Integrated Manufacturing Syetems, Edited by W. Dale Compton, NAE Press, 1988, pp167-199.

Wirtz, A., (1988), "Vectorial Tolerancing", Report from Neu-Technikum Buchs, CH 9470 Buchs, Gallen, Switzerland.

Wirtz, A. (1991), "Vector Tolerancing for Production Quality Control and Functional Analysis in Design", Proceedings of CIRP Intl. Working Seminar on Computer-Aided Tolerancing, Ham, I and Lehtihet, E. A., editors, Penn State, May 16-17, 1991.

APPLICATION OF QUALITY FUNCTION DEPLOYMENT (QFD) IN CONCURRENT ENGINEERING

Larry R. Smith*

SUMMARY

The technique that allows definition of customer requirements and carry them throughout the product and process design and development came to be known as Quality Function Deployment. Quality Function Deployment results when the cross-functional team compares, in matrix form, the customer driven definition of a good product to the planned activity elements of a program.

Quality Function Deployment (QFD) has been successfully applied to a number of specific designs to bring in the customer's requirements and wants into the designs from the very beginning of the conceptual design of products and processes. Although QFD dates back to the 1970's, its usefulness to the current concurrent or simultaneous engineering practices is well recognized only recently. The application of QFD is illustrated through two Ford Motor Company cases described in this paper: Case 1: Variable Assist Power Steering (VAPS): Use of QFD to formulate an Engineering Specification for a new product, and Case 2: Ford's 1991 CIII Power Steering Pump: Use of QFD in the Product Development Process. These cases are briefly presented to illustrate how QFD has become an available tool to formulate specifications from customer requirements and wants, and how a systematic phased approach has been implemented with QFD ideas adopted to the product development process.

QFD has become an established technique for defining the specific customer requirements at the system, subsystem, component and individual part levels.

* The author is the Manager of the Quality Strategy and Planning Department for Powertrain Operations, Ford Motor Company, Dearborn, MI

INTRODUCTION

Concurrent engineering deals with simultaneous (also called, concurrent) design and development of product and manufacturing processes. QFD has been applied as a means to deploy quality throughout the evolution of the product and manufacturing process design. QFD has proven to be particularly effective in breaking the traditional adversarial relationships between design and manufacturing. QFD has been found to foster inter-functional team building in carrying the voice of the customer throughout the organization. At Ford, the use of QFD has been a well established practice, where inter-functional teams use QFD to define and convey the customer requirements and preferences.

In a search for methods to define customer requirements, Ford came across Quality Function Deployment which Toyota had used to successfully identify and solve the excessive auto body rust problems it faced during the late 1970's. Ford engineers visited Japan to investigate QFD and to understand its successful applications. Among those who pioneered this investigation were Larry Sullivan of ASI (Formerly, a Ford Manager) and Don Clausing, MIT (Formerly a design engineer at Xerox Corporation). Ford engaged the services of leading QFD proponents such as Yoji Akao and Akashi Fukuhara and utilized organizations such as ASI and GOAL/QPC to conduct extensive training of its engineers.

Initial pilot QFD projects at Ford started in 1985. In 1987, QFD was defined as the Concept to Customer (C-T-C) process tool for use in product development, and the QFD subcommittee of C-T-C education and training was formed. In 1988, a QFD implementation strategy study group was created with membership representation from all activities within North American Automotive Operations (NAAO). Starting with the humble beginnings in 1985, the total number of Ford personnel trained and the number of projects has grown to 5,400 and 400 respectively by 1991.

The application of QFD at Ford for up-front engineering is at different levels: Product level addresses the entire vehicle; System level deals with major systems such as body, powertrain, chassis; Sub-system level tackles interior, exterior, engine, transmission, axles, brakes, suspension; Component level and Individual part level address to more and more detailed levels. QFD applications involve (1) Product Planning phases of vehicle development where activities of competitive benchmarking and identification of specific quality targets are performed; (2) Parts Deployment where concept selection and identification of significant parts and their characteristics are defined; (3) Process Planning where concept selection for significant operations and processes is done.; and (4) Production Planning where production control and maintenance requirements, mistake-proofing, and education and training of production personnel are performed.

In this paper two case studies are briefly described to illustrate how QFD may be used to: (1) to formulate product specifications and (2) to conduct product development process.

BACKGROUND

"Product standards or specifications are not always thorough, raw materials specifications are not always thorough and tolerance limits in drawings are not always thorough. (Therefore) I do not believe in product standards or quality specifications, raw material specifications, tolerance limits in drawings, measurement instruments, or chemical analysis. (And) this has been my key motto to be speaking (questioning) the very fundamentals of quality control". So wrote Kaoru Ishikawa, a well known Quality Consultant and Organizer (Ishikawa, 1984). He asked the crucial question: *how exactly does one improve the quality of a product or service?*

In Japan, quality control has progressed through three stages of development: (1) Inspection oriented quality assurance, (2) process control oriented quality assurance, and (3) product development quality assurance (Ishikawa, 1981). Ishikawa was one of the first to realize that quality could not be assured through the sole use of inspection. Although inspection has been used to weed out defective products, it also produces undesirable side effects: decreased productivity and increased cost. In the early 1950's, Ishikawa along with Deming, Juran, and others spearheaded the use of process control to improve upon the deficiencies of inspection alone. Initially, the results of process control methods were encouraging. In most implementations, the number of defective products decreased as process control improved. Processes became 'capable' of manufacturing parts to print. (Process capability as measured by C_{pk}, was maintained between 1.33 and 1.67. Refer to texts on statistical quality and process control for explanation of C_{pk}). Nevertheless, Ishikawa observed that many customers were unhappy with certain products, even though these products met all print and specification requirements and the processes that produced them had proper capabilities. After several attempts to resolve this dilemma, Ishikawa concluded that quality assurance must be applied in the new product development phase itself. To understand what quality means to a designer, he began asking a standard question during his consulting assignments, "When I ask the designer what is a good car, what is a good refrigerator, and what is a good synthetic fiber, most of them cannot answer. It is obvious that they cannot produce good products (Ishikawa, 1977)." Furthermore, he observed that those companies that designed quality into their products tended to greater customer satisfaction and consequently became more profitable. Ishikawa called this type of quality control MMK Quality Control, where MMK stands for "Mokatte Mokatte Komaru," or "profitable to a surprising extent", (Ishikawa, 1977). This act of defining quality in terms of customer requirements and then designing it into the product launched the next revolution in quality assurance.

The technique that allows definition of customer requirements and carries it throughout the process of product and process design and development came to be known as Quality Function Deployment. In Japanese, the word "deployment" refers to an extension or broadening of activity (Akao, and Kogure, 1983). Thus, Quality Function Deployment involves extending or broadening quality within a particular activity where an activity may involve anything from product planning to production assembly. It is accomplished through the creation of a matrix of customer requirements (quality table) versus planned activity elements, now popularly known as the 'House of Quality' (Hauser and Clausing, 1988). The first use of quality table, can be traced back to the Kobe Shipyard of Mitsubishi Heavy Industries in 1972, where it was found to be very effective when developed with input from a cross-functional team (Ishikawa, 1977).

QFD begins by considering the product or service from a customer perspective, defining what the customer would like if the product or service were ideal. These are the true quality characteristics which, if deployed through the product development cycle, result in a product or service which the customer is proud to own and will brag about to friends. Often these true quality characteristics cannot be directly measured (for example, a customer may desire a vehicle with "comfortable steering," a true quality characteristic). For this reason, it is necessary to relate true quality characteristics to substitute quality characteristics which can be readily measured. The resulting logic is typically several levels deep. (Among other things, comfortable steering is a function of steering wheel effort. Steering wheel effort is a function of pump flow capacity and several other variables. Pump flow capacity is partially a result of internal leakage due to clearance between the rotor vane and the pressure plate. Thus a rotor vane to pressure plate clearance is one substitute quality characteristic of comfortable steering). The quality table moves from the "world of the customer' to the "world of the engineer" by mapping the relationship between true and substitute quality characteristics. This map provides a customer driven definition of a "good product."

Basic Methodology in Applying QFD

Although QFD has evolved into different formats, depending on the application and scope, there are some basic steps that can be identified. The methodology presented by Clausing and Hauser, 1987 is perhaps the most straight forward and hence used to summarized the steps below.

(1) Collect raw data about the customer's requirements: Market research data, focused group interviews, on-site observations, demographics, and other market research means are used to collect what customers require. Objective expectations about performance, safety, durability, reliability, maintenance, repair as well as sensitivity about cost are among the data collected.

(2) Identify substitute characteristics and cross-reference them against each of the customer requirements: the product characteristics are the attributes of the product that influence the customer requirements positively or negatively. These are listed at the top of the 'House of Quality'.

(3) Obtain subjective ranking of your product with respect to those of the competitor's products: It is important to obtain customer perceptions on how he/she feels about the relative merits and demerits of your current product relative to those of your competitors.

(4) Conduct objective comparison of each of the product characteristics identified in (2) for yours and competitor's products: This produces quantitative assessment of the characteristics through actual testing, reverse engineering or performance evaluations. These are listed for yours and competitor's products at the bottom first horizontal row of the house of quality.

(5) Set targets: Specific objective targets are set for each product characteristics based on the inter-functional team's strategy for maximizing possibility for customer-satisfaction in relation to what the competitors are likely to do. This is a matter of strategic decision. Specific target rather than range are preferred because they force to establish the targets that must be achieved by design and manufacturing. Any tolerance or deviation around them must be done through robust design which maximizes the main function of the design despite the variabilities of the user environment and the manufacturing process.

This process can be repeated to deploy quality through successive houses for parts deployment, process planning and production. For each house the preceding house provides the input ('customer') requirements. For example, the part characteristic targets become the input to part deployment house, the parts characteristics become inputs to process planning house, and the key processes become the input to production requirements.

As seen above, QFD results when a cross-functional team compares the customer driven definition of a good product to planned activity elements of a program using the QFD matrix form. Often these activity elements must be modified to provide adequate coverage of customer needs. QFD is, therefore, a planning tool that identifies the critical items that must be addressed to achieve customer satisfaction. The QFD matrix may be augmented by other tables for purposes of providing information as to (1) how activity elements interact with one another to prioritize activity elements, and to benchmark activity elements against competition. QFD charts may also be cascaded in an interlocking series for the purpose of planning a sequential group of activities related to one overall program (Akao, Ishizuka, and Kamizawa, 1978; Akao, and Mizuno, 1985; and ASI, 1987).

The following two examples utilize the basic ideas of QFD as a means for deploying concurrent engineering. No attempt was made to duplicate any particular QFD format. The format evolved to reflect the specific problems under study. Akao, a leading Japanese authority on QFD states, "Generally speaking, the best method for quality deployment is not to play-by-the-book. Since the situation of each company differs, it is important to be resourceful in using the rules, and finding a method suitable for your situation (Akao, 1987)".

Case 1: Variable Assist Power Steering (VAPS): Use of QFD to formulate an Engineering Specification for a new product.

All too often, specifications are written late in the development process after the design has been detailed and just prior to product manufacture. This practice does not allow customer requirements to be reflected in the specifications. The specifications merely state the limitations of the product already designed and developed! Preferably, the specification should be first established from customer requirements and wants, which in turn should determine the product design concept and details. The Variable Assist Power Steering Valve for the Lincoln Continental was developed through this preferred method.

The Variable Assist Power Steering (VAPS) valve regulates the flow of power steering fluid between two gear spots, varying the level of power assist with vehicle speed. Although the concept was patented by Ford, the design and development work was sourced to an outside supplier as a "black box' item. It was necessary, in the summer of 1985, to write an Engineering Specification for this component. This specification would have a major impact on the component and was to become a part of the contractual agreement between the supplier and Ford Product Engineering. Because the VAPS valve was based on a new patent, its specification could not simply be copied from historical documents. We decided to ask a significant question: How could the specification of VAPS be written in a manner that assures customer satisfaction?

Quality Function Deployment was used as a tool to write the Engineering Specification (ES) for the VAPS valve. Considering the valve's function from a customer's perspective, a four level quality (QFD) table was constructed, a portion of which is shown in Figure 1. A series of proposed ES tests were designed and evaluated against this table so that the relationship between each ES test and customer requirements could be identified and defined. When this was first completed, it was evident that there were gaps in the resulting matrix; areas where customer needs were not being directly addressed. For this reason, tests such as thermal cycling and impulse were added to the proposed specification and other existing tests were modified. A number of areas were identified which would be dealt with in a separate specification written by the supplier to a sub-supplier.

A comparison of proposed tests to other ES tests in the power steering system was added to the bottom of the QFD matrix. This comparison was very useful in selecting target values for valve test requirements (some tests were modified based on this system comparison). This case proved the value of QFD in formulation of product specification from customer requirements and wants.

Those benefits of QFD are:

• The development test program becomes tailored to meet customer needs.

• The supplier understands how and why the specification tests were established. This becomes important later in the program. For example, when, after several late material changes, valves could not pass several ES tests, the QFD reasoning helped the problem solving team focus on design issues rather than debate the validity of Engineering Specifications.

• The QFD comparisons could be used as a tool to achieve system test uniformity. For instance, frequently, systems are broken into smaller components which become the responsibility of different design groups. Each of these groups may write specifications which address common system issues. However, this results in component tests which may differ from one another as was found in the case mentioned above where the tests devised by different groups differed in terms of required life, test temperature, and stress. The QFD approach has been found to have a great value in bringing these groups together for formulating specifications that are consistent with respect to common system issues.

Case 2: Ford's 1991 CIII Power Steering Pump: Use of QFD in the Product Development Process

QFD was used as an integral part of the product development process in the design program for Ford's 1991 CIII Power Steering Pump shown in Figure 2. In this application, QFD was used to integrate various quality tools into a planned program to build quality into the part during the design stage. The strategy consisted of four phases (diagrammed in Figure 3 as: Phase I: Competitive Comparison, Phase II: Engineering Program Development, Phase III: Significant Characteristic Selection, and Phase IV: Dimensional Control Plan), each of which will be described in detail below.

Phase I, partially shown in Figure 4 and created for the purpose of competitive comparison, consists of a very general description of pump quality versus a list of design criteria which can be used to evaluate one pump against another. The resulting chart was used to establish performance objectives for the new pump, and highlight competitive advantages to Ford management. Because a typical customer may not be aware that a power steering pump exists, warranty and internal customer feedback were used to determine customer wants and priorities; engineering tests and pump tear-downs were used for technical benchmarking. Relationships between design criteria are shown at the top of the QFD matrix.

Phase II was used to plan the Engineering Program and is partially shown in Figure 5 and 6. For this phase, a quality table was first constructed to provide a customer driven definition of a good pump. A significant teaching and learning benefit occurred as the group, consisting of engineers with varying degrees of experience, mapped the "world of the customer" to the 'world of the engineer." In effect, each person was able to share their knowledge with the rest of the team, and the ensuing conversation generated intense interest. Phase II was completed by laying out customer to program elements in matrix form. This matrix immediately identified areas where customers had needs that were not being directly addressed during the development process. Therefore, additional test elements were added to the program for noise analysis, seal leakage, and thermal cycling. The program also became more efficient as redundant test elements were formally identified and eliminated.

Phase II was also used to integrate or blend various quality tools into one comprehensive program. For instance, a portion of the quality table dealt with what the internal customer wants for ease of assembly from the parts manufactured. This portion utilized Design for Assembly techniques to move from the customer want to the engineering reality. In addition, specific areas of customer

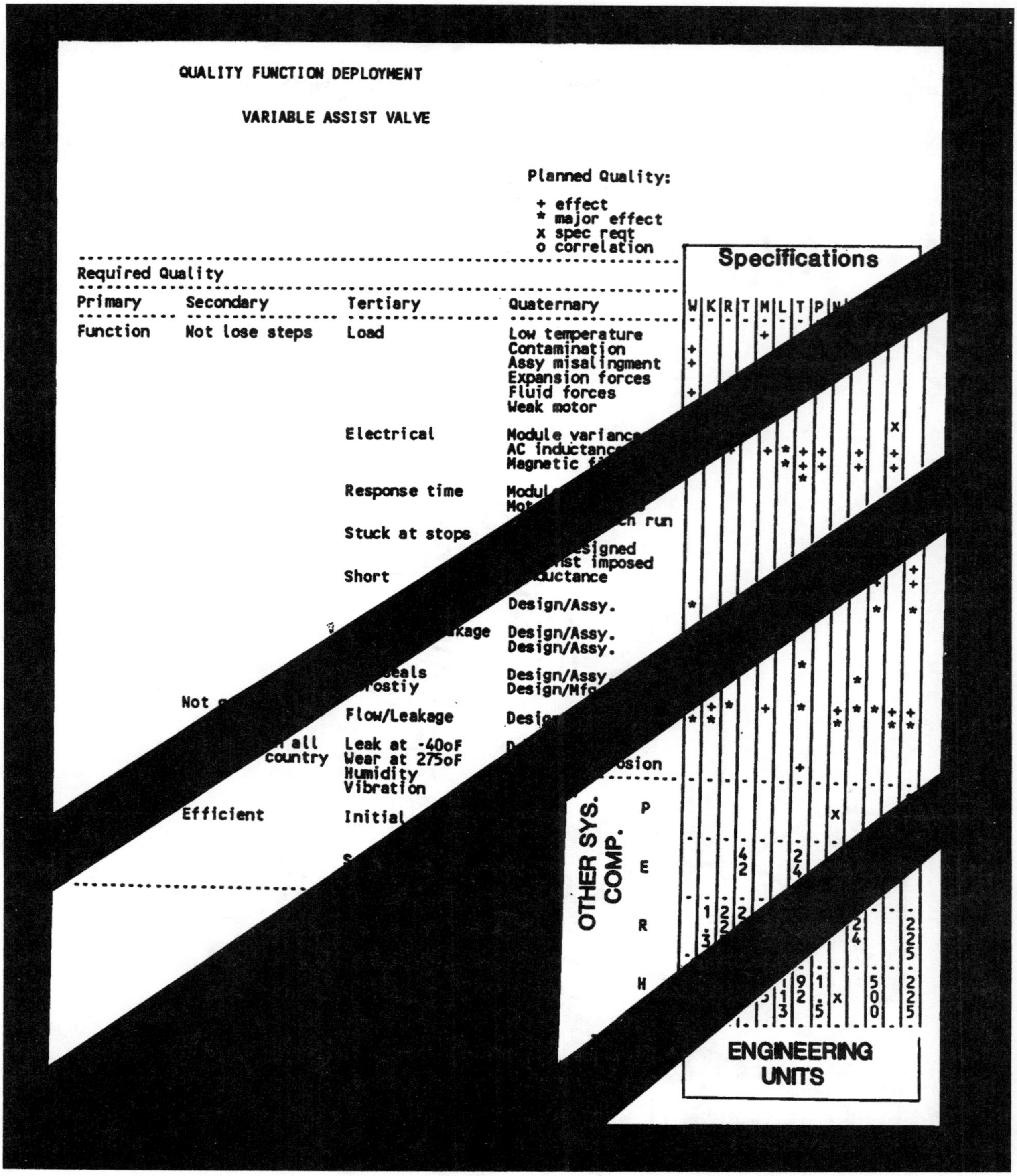

FIGURE 1: A Four Level QFD Table for Variable Assist Valve

C-III POWER STEERING PUMP

QFD Program

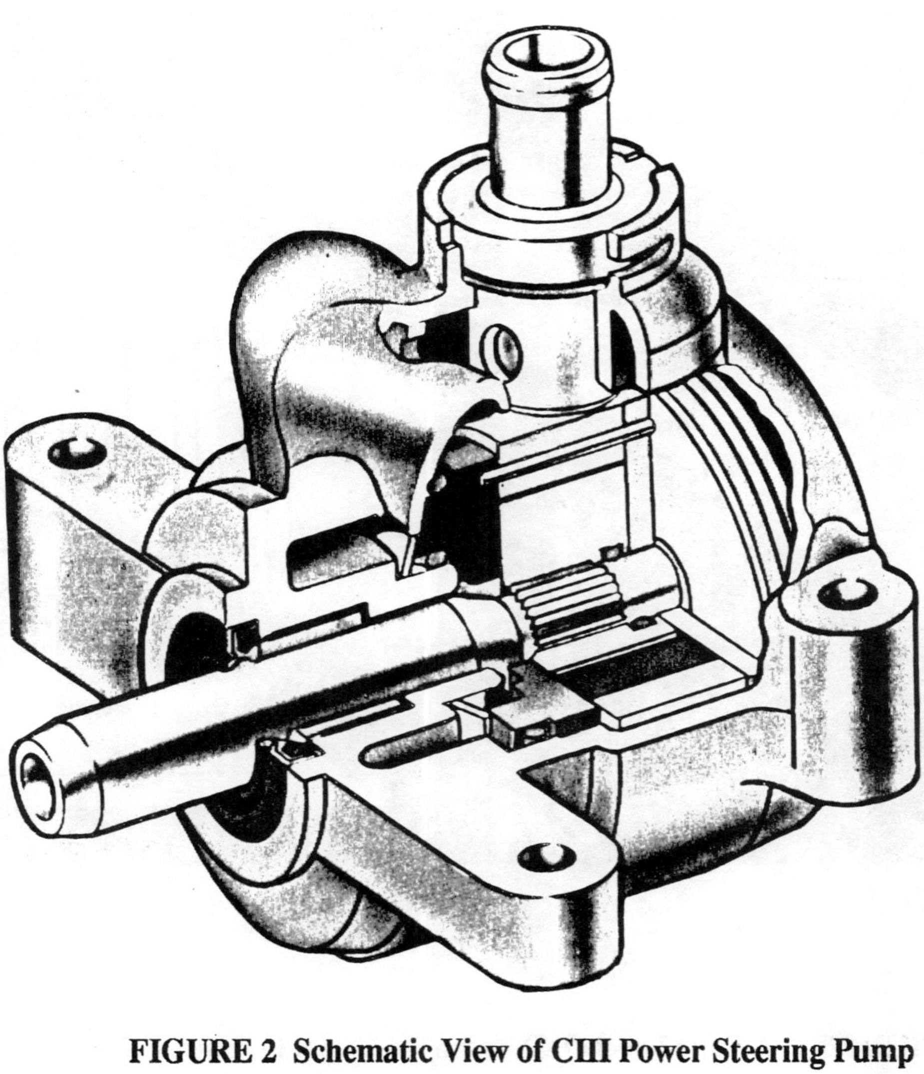

FIGURE 2 Schematic View of CIII Power Steering Pump

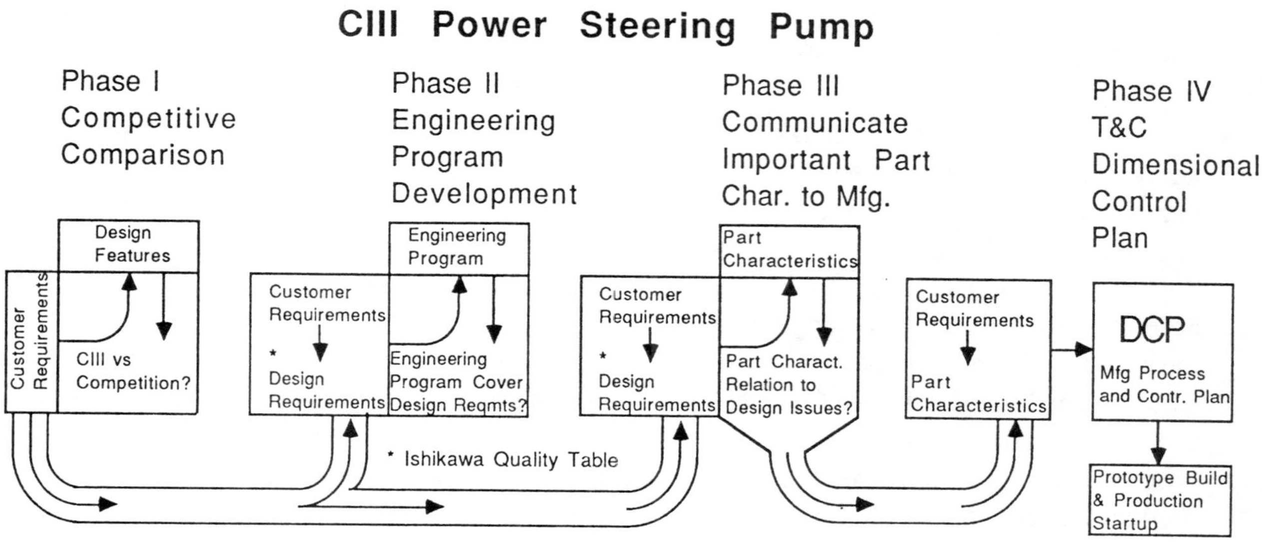

Figure 3 Four Phase QFD Strategy Applied to CIII Power Steering Pump

concern were identified for further work using Design for Experiments and/or Computer Aided Engineering.

As a part of the development program, the items in the main pumping group (cam, rotor, vanes, and valve plate) were computer simulated for both the new pump and existing competition. QFD was used to identify and prioritize those aspects of the pump that needed to be analyzed through computer simulation. For example, the quality table indicated that the structure born noise is reduced when pressure pulse magnitude is reduced. A simulation of CIII pump pressure pulse magnitude versus the competition indicated the final design of the new pump was "Best In Class." Many other useful by-products of the simulation have been developed including generation of X-Y coordinates for manufacture of prototype cams, and the prediction of clamping force required to retain the cam during production grinding.

The design of an improved supercharger circuit is an example of how experimental usage was planned into the product development program. Per the quality table, supercharger pressure is related to airborne noise (whine and hiss) and a customer demand for a quiet pump. Experiments coupled with engineering knowledge allowed the engineer to arrive at a design with proper performance in a relatively short period of time.

The Phase II QFD provided a method of creating a design/manufacturing program that addressed customer requirements in a thorough, cost effective, and time efficient manner. Many benefits of the early QFD work were not realized until much later in the program. During a discussion of pump packaging associated with the creation of the quality table, it was mentioned that several dollars per vehicle would have been saved on one past program if the current pump could run in reverse. This was documented as a desired feature (see Figure 6). The team then determined that the pump could be manufactured to run clockwise or counterclockwise if the manufacturing process was designed to interchange four parts and built this flexibility into the program before long lead tooling was ordered. A year later, when this feature became a necessary part of Ford's Zeta engine program, the manufacturing flexibility had already been established. In a few hours the team finalized details that, at that time, would have taken months of cost related study in a traditional program that did not focus or anticipate customer needs.

Phase III of the QFD (partially shown in Figure 7) was created to communicate important part characteristics to manufacturing and utilized the same quality table that provided customer focus in the engineering program. In the Phase III matrix, print characteristics were related to the quality table to provide a "map" of how the drawing dimensions and notes relate to external and internal customers. A separate matrix was created for each part drawing. Creation of this phase generated a great deal of discussion and information sharing between the manufacturing and design engineers. The manufacturing engineers better understood why the product functions as it does and how specific production variations affect product performance: the designers gained a much deeper appreciation for manufacturing process capability and controls. An indirect benefit of these discussion was the creation of a tight cohesive design/manufacturing team which worked together to quickly solve both QFD and non-QFD related issues. Through an improved understanding of customer sensitive characteristics, the manufacturing engineers were able to design the process to reduce variation in selective areas important for customer satisfaction early in the product program. The Phase III documents are also a good reference tool for future changes and control system changes.

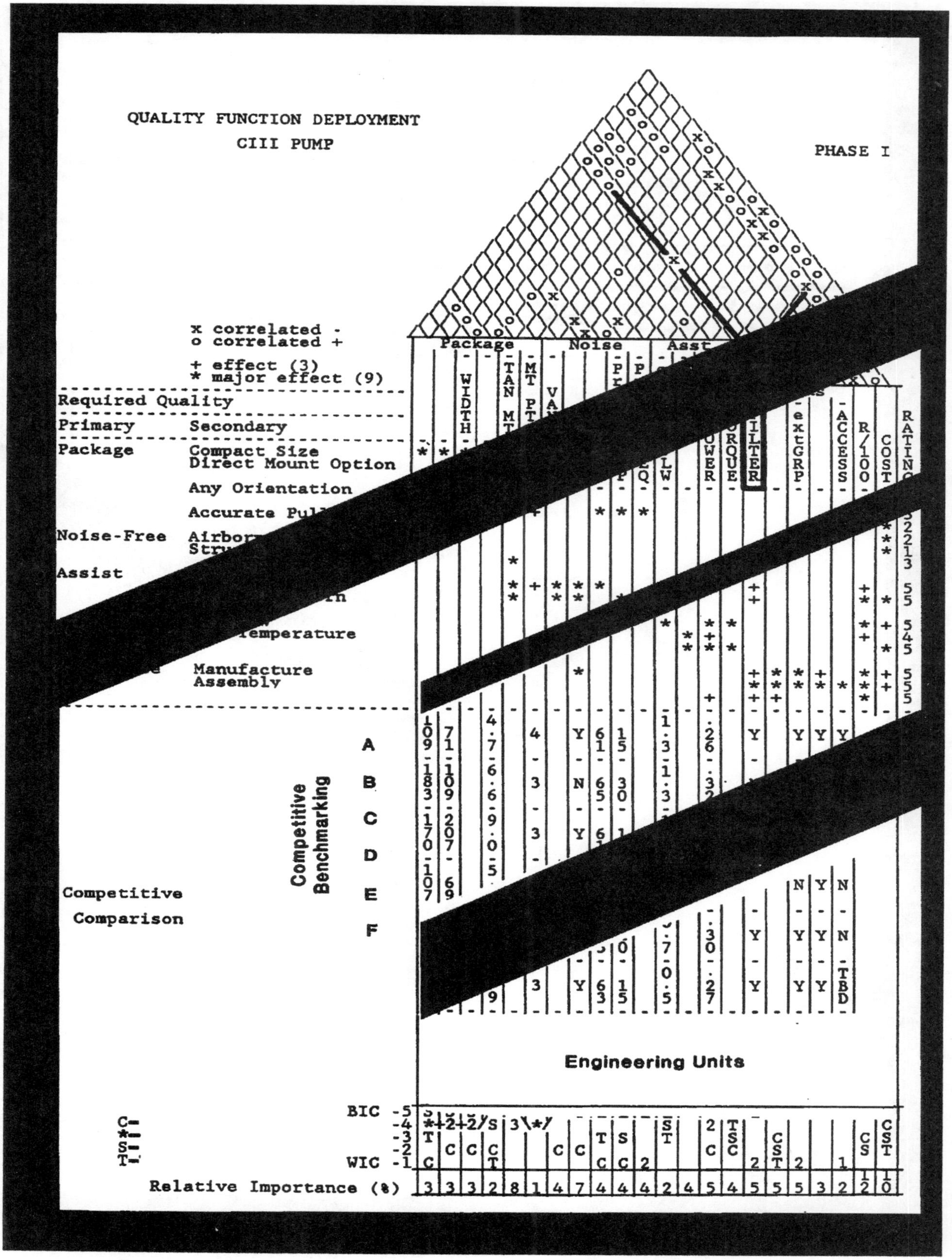

FIGURE 4 Phase I Quality Function Deployment for CIII Pump

Establish Engineering Program

CIII QFD PHASE II

ISHIKAWA QUALITY TABLE

Page 1 of 2

+ Minor Effect
* Major Effect

Primary	Secondary	Tertiary	Quaternary	Quintary	QUIET RM	VEH NVH	DUR/FLET	HI SPEED	PR CYCLE	21 CYCLE	edSEISURE	cmWIPEOUT	hyIMPACT	MF AUDIT	airPURGE	VLVrspTM	CAP FLOW	COLD SIM	WEIGHT	COMP SIM	DOE
NOISE-FREE AIRBORN		WHINE-CAVITATION	PORT DESIGN	TIMING (ANGLE)	*	*		+	+											*	*
				DEPTH (AXIAL)	*	*		+	+											*	*
				LENGTH (RADIAL)	*	*		+	+											*	*
			SUPERCHARGE	Hv < (1.6)Hi	*	*	+	*	+			+			+						*
			NOISE NOTCH		*	*	+	*	+											*	*
				DENSITY			+	+	+	+		+									+
				VISCOSITY			+	+	+	+		+						*		*	+
							+	+	+	+		+								*	[illegible]
				LUBRICITY			*	*	+	+		+							[illegible]	[illegible]	[illegible]
				[illegible]			*	+	+	+		+				[illegible]	[illegible]	[illegible]	[illegible]	[illegible]	[illegible]
		MECHANICAL NOISE	CLEARANCE	VANE/SLOT	*	*	*	*								[illegible]	[illegible]	[illegible]	[illegible]	[illegible]	[illegible]
			ENGAGEMENT	VANE HGT/ROTOR	*	*	+			+	+		[illegible]	[illegible]	[illegible]	[illegible]	[illegible]	[illegible]	[illegible]	*	*
			MATERIAL	HOUSING	*	*	+		*			[illegible]	[illegible]	[illegible]	[illegible]	[illegible]	[illegible]	[illegible]		*	*
			ACCURACY		*	+	+	+	+	[illegible]	[illegible]	[illegible]	[illegible]	[illegible]							
				END PLATE	*	+			[illegible]	[illegible]	[illegible]	[illegible]	[illegible]								
				RESONANCE FREQ	+	+	[illegible]	[illegible]	[illegible]	[illegible]	[illegible]	[illegible]								+	
					+	[illegible]	[illegible]	[illegible]	[illegible]	[illegible]	[illegible]						*			+	+
		HISS/SWISH	RELIEF VALVE		[illegible]	[illegible]	[illegible]	[illegible]	[illegible]	[illegible]										+	+
			SUPERCHARGE		[illegible]	[illegible]	[illegible]	[illegible]													
				HOUS[illegible]	[illegible]	[illegible]															
			RELIEF VALVE		[illegible]																
		MOAN	NOISE NOTC[illegible]																		
																					*
																					+
																					+

FIGURE 5 Phase II Quality Function Deployment for CIII: Establish Engineering Program

Phase IV was used to finalize the manufacturing process laws and to establish a control plan. This phase utilized an existing procedure, the Dimensional Control Plan (DCP), which was already well established in the Ford culture [9,10]. Although other companies may use QFD to accomplish what DCP does, it is better to use relevant portions of QFD as a tool to blend existing business elements together to form a comprehensive program with customer focus.

Use of Quality Function Deployment in the product development process has resulted in the following benefits:

- A better understanding of product dimensions and performance parameters in relation to competition and customer needs.
- A more effective engineering program with improved development tests and enhanced product flexibility in key areas of customer concern.
- Creation of an organized test and quality control program which assures the meeting of customer needs.
- The introduction of a mechanism to build strong cross-functional teams which pool resources to achieve the common goal of customer satisfaction. This commitment from one team member to another carries over into many other common activities that are not QFD related.
- Strong teaching and learning benefits for all team members.
- Documentation of design/manufacturing product knowledge for future reference and use.

QFD can provide the necessary organization and teamwork for effective simultaneous (concurrent) engineering with **customer focus**, an element that a concurrent engineering program might otherwise lack (Schaal, and Slabey, 1991).

REFERENCES

Akao, Y. Ishizuka, I., and Kamizawa, N., (1978), "Quality Evolution System and FMEA," Proceedings of the International Conference on Quality Control, !978, Tokyo, pp. B4.19-B4.28.

Akao, Y., and Kogure, M., (1983), "Quality Function Deployment and CWQC in Japan", Quality Progress, vol. 16, no. 10, Oct., 1983, pp. 25-29.

Akao, Y., and Mizuno, K., (1985) " Quality Function Evolution," Published by JUSA, Translated by Ford Motor Company, May, 1985 from Japanese text, pp. 1-255.

Akao, Y., (1987), "Quality Deployment," A series of articles, Growth Opportunities Alliance of Lawrence, Inc., Lawrence, MA., 1987, pp. 1-1 to 12-12.

Akao, Y., (1987), "Quality Deployment": A series of articles (Translated by Glen Mazur), GOAL/QPC, Methuen, MA.

American Supplier Institute Report, (1987), "Quality Function Deployment-A collection of Case Studies," Section, I-VIII., ASI, Dearborn, MI.

Hauser, J. R. and Clausing, D. (1988), "The House of Quality", HBR, May-June, 1988, pp. 63-73.

Ishikawa, K., (1977), "Quality Analysis", 1977 ASQC Technical Conference Transactions, Philadelphia, pp. 423-429.

Ishikawa, K., (1981), "Company Wide Quality Control-Revolution-Management," 1981 ASQC Quality Congress Transactions, San Francisco, pp.124-131.

Ishikawa, K., (1984), "Quality and Standardization: A Program for Economic Success", Quality

CIII QFD PHASE II

ISHIKAWA QUALITY TABLE

Page 2 of 2

+ Minor Effect
* Major Effect

Primary	Secondary	Tertiary	Quaternary	Quintary	LN/HG/WD	WEIGHT	FEATURE	MF AUDIT	DUR/FLET	TM CYCLE	COLD SIM	DUST/COR	HI SPEED	PR CYCLE	21 CYCLE	hyIMPACT	HS BURST	CA.P FLOW	airPURGE	COMP SIM	DOE
PACKAGE	COMPACT SIZE	RESERVOIR			*	*	+		*	+		+							*		
		PUMP			*	*	+		*			+				+	+				
	MOUNTING	RADIAL	WITH SUPPORT				+														
			W/O SUPPORT				+														
		TANGENTIAL	THREE POINT	HOLE & SLOTS			+														
				DOWEL			+														
			FOUR POINT	HOLE & SLOTS			+														
				DOWEL			*														
	OPTIONS	REMOTE RES					*														
		ANY ORIENTATION					*														
		RUN REVERSE	FEW CHANGES				*														
		DROOPER PIN					[illegible]	[illegible]	[illegible]												
	FITTING	INLET POSITION	STRAIGHT		[illegible]	[illegible]	[illegible]	[illegible]	[illegible]												
			ANY ANGLE																		
		OUTLET POSITION	STRAIGHT																		
			ANY ANGLE																		
	PULLEY	BOLT-ON																			
		PRESS-ON																			
LEAK-FREE	HOUSING	POROSITY	C[illegible]																		
			[illegible]LE																		

FIGURE 6 Phase II Quality Function Deployment CIII Pump: Ishikawa Quality Table

CIII QFD PHASE III

INSERT - PUMP CAM (FINISH)

CIII PUMP QUALITY MATRIX

Page 7 of 22

\+ Minor Effect
* Major Effect

Primary	Secondary	Tertiary	Quaternary	Quintary	22.5	9 DEGREES	CENTER CONTOUR NOTE	PARALLEL 0.005	18.995-19.005	FLATNESS 0.005
NOISE-FREE	AIRBORN	WHINE-CAVITATION	PORT DESIGN		*	*				
			SUPERCHARGE	Hv < (1.6)Hi						
			NOISE NOTCH	PRESS WAVE FORM						
			CLEARANCE							
				VANE HT/ROTOR/						
			MATERIAL	HOUSING						
			ACCURACY	CAM PR						
				END						
		HISS/SWISH								
			SUPERCHARG							
			MATERI							
		WHISTLE	RE							
	STRUCTURE	MOAN		CONTAM						
				BURR						
		RELIEF VALVE								
		RADIAL								

FIGURE 7 Phase III Quality Function Deployment CIII Pump Quality Matrix

Progress, vol. 17, no. 1, Jan. 1984, pp. 16-20.
King, R., (1987), " Better Designs in Half the Time": Implementing QFD in America, GOAL/QPC, Methuen, MA.
Schaal, H. F. and Slabey, W. R., (1991), "Implementing QFD at Ford", Ford Report.
Sullivan, L. P., (1988), "Quality Function Deployment", Quality Progress, vol XXI, no. 6, June, 1988.

THE VISION OF SIX SIGMA QUALITY

Producibility Measurement and Analysis

Mikel J. Harry*

SUMMARY

The purpose of this paper is to provide the reader with a conceptual window from which to view the universe of producibility assessment in the context of Motorola's Six Sigma initiative. Although the focus is on product design, we shall also investigate its relevance to process design and manufacturing control. More specifically, we will develop a fundamental understanding of: a) why producibility must be studied prior to the release of a design for production; b) the role that process, material, and component variation plays with respect to producibility; c) the specific metrics used to study producibility a priori; and d) the methodology necessary for applying the producibility metrics.

In the context of this paper, the first step toward the quantitative assessment of producibility entails the capture of process, material, and component data. Next, the data are statistically summarized into indices of capability. Following this, the summary information would be transferred to a computer data file. During the course of product design simulation, the file would be periodically accessed by the designer so as to develop the most cost effective combination of parts, material, processes, etc. Naturally, if actual capability data does not exist, the designer would be forced to postulate the state-of-affairs. Once the baseline simulations and engineering studies are complete, the projected parameter yields would be multiplied and subsequently normalized. In turn, the normalized yields would be used to create an overall statement of producibility.

Implementation of producibility involves essentially, the characterization strategy which consists of 4 primary phases: 1) product parameter definition, 2) product parameter analyses, 3) process parameter optimization, and 4) process parameter control. In this paper, these are discussed in view of providing a sound foundation for producibility measurement and its relationship to the Six Sigma initiative at Motorola.

* The author is the Director of Six Sigma Research Institute and a Senior Member of Technical Staff Six Sigma Research Institute, Motorola University, Motorola, Inc., Schaumberg, IL.

INTRODUCTION

The Motivating Force for Change

The customers which form the enormous base of today's world market are sending a clear and undeniable message to Corporate America. They are demanding, without compromise, higher levels of product quality, at a lower cost, greater responsiveness, and added value. Motorola Inc. is a company which has heard the message and risen to the challenge. According to USA Today, "Motorola engineered one of the most dramatic [business] comebacks of this decade." Hillkirk, J.(1989).

This effort was formally acknowledged by President Reagan at the White House in 1988, by presenting the first *Malcolm Baldridge National Quality Award* to Motorola, Inc. At the heart of Motorola's success story is a single, overriding objective. Namely, the objective is Customer Satisfaction. This means satisfying the various technical, performance, schedule, and cost expectations of the customer. In turn, this objective drives the need for *Six Sigma* and *Total Quality Management (TQM).*

The Nature of Six Sigma Quality

For several years, Motorola Inc. has been pursuing an extremely aggressive quality goal-the implementation of Six Sigma. As many already know, this goal translates to 2 non-conformities-per-billion-opportunities for nonconformance (npbo). In light of normal disturbances in process centering, the goal is altered to 3.4 nonconformities-per-million-opportunities for nonconformance (npmo). To better understand the underlying logic, we must turn to mathematical statistics.

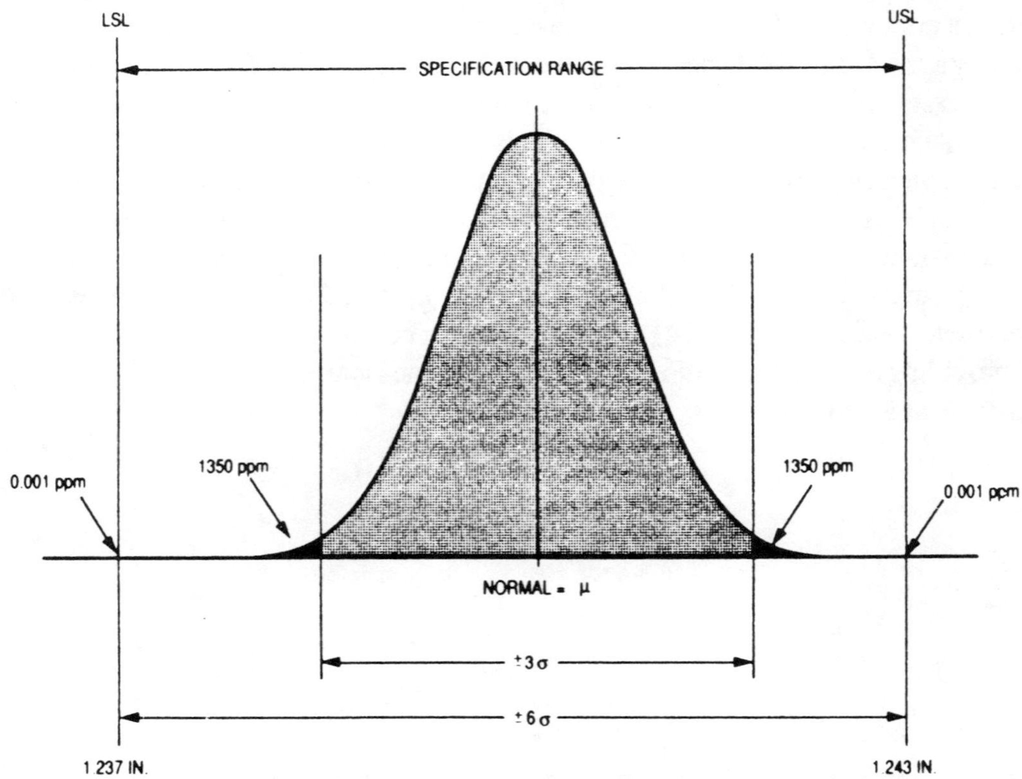

Figure 1. Centered Six Sigma Distribution

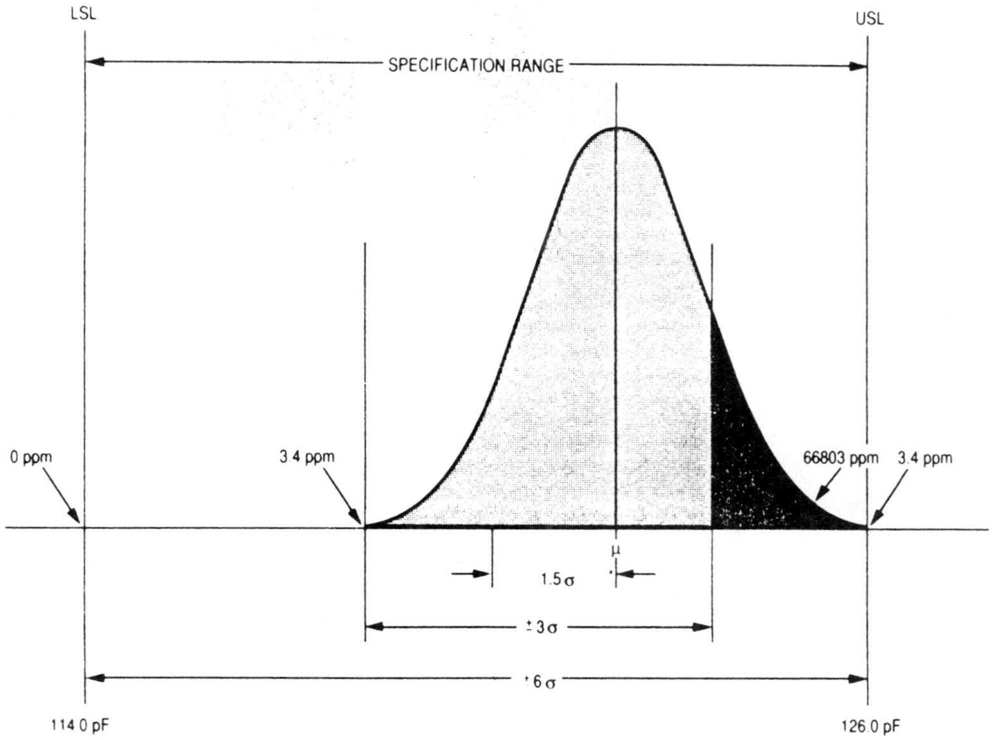

Figure 2. Six Sigma Distribution with a Mean Off-Set of 1.5σ

If we were to consider a random performance variable (X) in the context of a bilateral engineering tolerance, the performance goal of 2 npbo can also be described by a $\pm 6\sigma$ distribution, assuming that $X \approx NID(\mu,\sigma)$. To account for dynamic perturbations in process centering, the performance of X is decreased via a compensatory off-set in the mean (μ) such that $\delta = 1.5\sigma$. Hence, the adjusted goal of 3.4 npmo. For the reader's convenience, the latter two conditions are illustrated in figures 1 and 2 respectively. For comprehensive discussion of Motorola's Six Sigma initiative and the related statistical details, the reader is directed to Harry (1986), as well as Harry and Lawson (1988).

The Nature of Total Quality Management

We must all recognize that the TQM philosophy provides a comprehensive way to improve quality by examining the way work gets done in a systematic, integrated, consistent, organization-wide perspective DoD (1989). This particular philosophy steers us to a new way of doing business. In fact, the primary focus is on such things as: a) continuous improvement, b) cross-functional team involvement, c) modern process control techniques, and d) the effects of variation on processes, just to mention a few. The Participative Management Program (PMP) and Six Sigma initiative at Motorola Inc. embodies not only the fundamental precepts of TQM, but also provides a vehicle for its successful implementation.

Rising to the TQM/Six Sigma challenge is not an easy task. In fact, one of the biggest deterrents to meeting this challenge is the reduction and ultimate elimination of product variation. When variation acts upon a product, the results are often catastrophic in terms of reliability, quality, performance, schedule, and costs. Although the task of eradicating variation is a tough one, there are a number of things which can be done. To this end, Motorola Inc. has structured several key initiatives. Those highly interactive initiatives are inclusive of, but not necessarily limited to :

Design
•Design to Standard Parts
•Design to Standard Materials
•Robust Design
•Design for Assembly
•Design for Reliability
•Design for Simplicity

Process
•Short-Cycle Manufacturing
•Process Characterization
•Process Standardization
•Statistical Process Control

Material and Components
•Part Standardization
•Supplier SPC
•Supplier Certification
•Material Requirements Planning

As may be apparent, the aforementioned initiatives have been devised to achieve Six Sigma Synergy. This is to say that each of the Six Sigma initiatives and related tools are cross-functional by nature, therein allowing the whole to be greater than the sum of the parts.

The Issue of Producibility

Also in recent times, we have experienced an enormous explosion of technology. As a result, a large proportion of today's product designers have been liberated from many types of material and component related constraints. Because of this technological liberation, we see many familiar products which operate faster, have more features, occupy less space, and in some instances, even cost less than their predecessors.

The overriding manufacturing implication of the so called: "technology boom" is quite clear—for every incremental increase in design complexity and sophistication there must be a corresponding increase in producibility; otherwise the manufacturer will not remain competitive. In many factories, producibility has become a major business issue. It is often the key to economic success or catastrophe.

The purpose of this paper is to provide the reader with a conceptual window from which to view the universe of producibility assessment in the context of Motorola's Six Sigma initiative [note #1]. Although the focus is on product design, we shall also investigate its relevance to process design and manufacturing control. More specifically, we will develop a fundamental understanding of: a) why producibility must be studied prior to the release of a design for production; b) the role that process, material, and component variation plays with respect to producibility; c) the specific metrics used to study producibility a priori; and d) the methodology necessary for applying the producibility metrics.

THE BROAD UNDERSTANDING OF PRODUCIBILITY

The Definition of Producibility

As we are all aware, the notion of "producibility" encompasses a myriad of seemingly complex topics. However, before we venture into the forest of technical detail, so to speak, let us gain an overall picture of what the term "producibility" actually refers to. In this manner, we can create a conceptual topology of the forest.

At a conference sponsored by the Department of the Navy, the term, "producibility" was broadly defined as "...as measure of the relative ease of manufacturing a product (U. S. Navy, 1989)" Hence, we may say that producibility is concerned with how easily a design concept can be transitioned from development on through the production of a quality product at the lowest possible cost, with particular emphasis on quality. As may be apparent, this understanding of producibility ultimately leads to two fundamental issues—measurement and the viability of manufacture.

The Issue of Measurement

Perhaps the emphasis on measurement can be better understood by the message Lord Kelvin extended in 1891 (Kelvin, 1891). He stated:

...When you can measure what you are speaking about and express it in numbers, you know something about it, but when you cannot express it in numbers, your knowledge is of a meager and unsatisfactory kind.

Hence, we may now assert that the issue of producibility or "manufacturing ease" must be quantifiable; i.e., measurable, if it is to be satisfactorily studied and subsequently used as a basis for decision making and design benchmarking [Note #2]. After all, if a producer can not measure the extent to which a product can be transitioned from development on through production, it is virtually certain that their knowledge of producibility is of a "meager and unsatisfactory kind." With this in mind, we may say now that:

- If the producer does not know much about producibility, it is virtually certain that they are not in control of it.
- If the producer is not in control, then the producer and customer are surely at the mercy of chance.
- If both the producer and customer are at the mercy of chance, it is virtually certain neither will be satisfied.
- If the customer is not satisfied, it is virtually certain that the customer will seek satisfaction elsewhere.
- If the customer seeks satisfaction elsewhere, it is virtually certain that the producer will no longer do business with that particular customer.

As a consequence of this logic, we may conclude that the ability to measure producibility is of major concern and perhaps, the "yellow brick road" to manufacturing success. However, before we can discuss the specific engineering metrics used to study producibility, we must first articulate the second fundamental issue. In short, we must address the phrase "manufacturing ease."

The Issue of Manufacturing Viability

In order to clarify the issue of "manufacturing ease," we must consider the impact of product variation on producibility. To illustrate, let us consider a certain critical product characteristic, say Y_1. In this instance, we can let Y_1 be whatever comes to mind; e.g., the length of a mechanical part, voltage in a circuit, resistance of a material, etc.

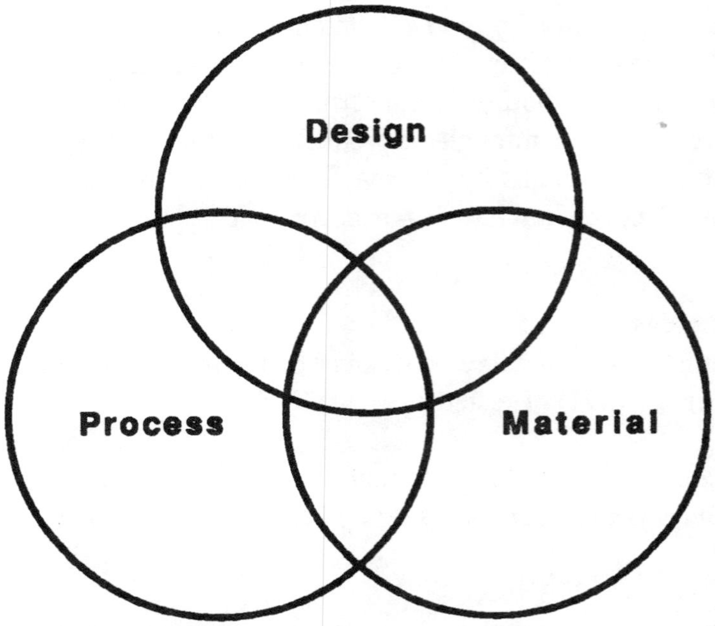

Figure 3. Relationship between the three Primary Sources of Variation

Let us suppose that Y_1 exhibits a large amount of variation to its relative performance standard. Under this condition, we may say that "ease of manufacture" is low when compared to the situation when the variation in Y_1 is small. Thus, as variation increases, producibility decreases because the probability of nonconformance increases. In turn, this forces additional rework, scrap, cycle time, cost, etc. As rework and scrap increase, direct labor increases. Obviously, this implies working harder, not smarter. Hence, the "ease" to which one can manufacture a product is diminished as variation increases. Therefore, we may say that variation and producibility are inversely related. But where does such variation come from?

As many of us are all too aware, product variation emanates from the underlying network of casual variables. Perhaps this understanding would be better expressed as, $Y = f(X_{1,...}X_n)$ where Y is some product characteristic and $(X_{1,...,}X_n)$ describes all of the variables in the cause system. Interestingly, the many variables within most any cause system can be classified into three primary sources of causation. Namely, these sources are: a) inadequate design margin, b) insufficient process control, and c) unstable material and components. The relationship among these three sources have been summarized in figure 3.

It is all too apparent how the independent effects of the numerous design, process, and material variables and their combinatorial interactions form the foundation of product quality. When the "vital few" effects are controlled such that variation is minimized, product quality is improved. As quality is improved, producibility is enhanced, for the reason given earlier. Hence, we can now better understand the interplay between product quality and producibility. Along this same line of reasoning, Stoll, Kumar, and Maas (date unknown) stated:

Producibility and quality are closely coupled because both are essentially the resultant or integrated effect of a myriad of diverse and complex interactions between the product and the manufacturing processes which produce it within a complex environment. These interactions involve all components of the manufacturing system. Inattention to these interactions during the design phase generally results in poor product quality, costly production problems, and a steady stream of engineering change notices (ECNs) during the production phase. Conversely, quality during the production phase is the result of a "producible design" created during the design phase. A producible design results from paying close attention to product/process interactions and seeking to minimize these interactions and their impact by designing the product and process as a coordinated system. In this sense, producibility can be thought of as "quality during the design phase."

Synopsis of the Issues

Thus far, we have determined that variation is the major enemy of producibility. We have also observed that we must describe variation numerically if we are to study and ultimately improve the ease by which products are manufactured. In particular, we must be able to quantitatively assess variation with respect to design margins, material/component specifications, and process control/capability. Only then can the issue of producibility be addressed in a practical and meaningful way. Obviously, if the producer can not express what they know in numbers, control can not be exerted and, if control can not be realized, the producer and customer are at the mercy of chance.

As should be readily apparent, if one is at the mercy of chance, prediction is not possible-which is the entire point of studying producibility a priori, or as some would say, "before-the-fact." Therefore, it may be concluded that when the amount of variation is large, relative to the performance standards, one's ability to predict the frequency of nonconformance diminishes. As a logical consequence, manufacturing viability is lost.

In light of the latter arguments, it may now be reasoned that the accurate prediction of producibility is highly dependent upon a measure of the interplay within and between the three circles depicted in figure 3. In addition, we may also say that the optimization of producibility during the design phase requires that we "design for producibility." This implies that the product designs will be relatively impervious, or as some would say, "robust" to natural, unavoidable sources of process, component, and material variation. In turn, this assumes that we have a quantitative knowledge of process, component, and material capabilities which is continually updated and renewed.

Thus, if we are to pursue manufacturing excellence, we must strive to numerically optimize product and process designs in accordance with known production capabilities. As may be apparent, this implies that we design products in the light of that variation which we know is inevitable rather than in darkness of chance. It is also to say that we must assess the manufacturing viability of our designs before releasing them for production. It should go without saying that the synergistic effect of designing for manufacturability and producibility assessment react in such a fashion that the whole is far greater than the sum of the parts—no pun intended.

THE BASIS FOR PRODUCIBILITY METRICS

The Producibility Game

At this point in our discussion, one might well respond: "so far, all of this sounds good in theory, but how does it translate to a better understanding of producibility ?" Well, in order to fully address this question, we must first overview some key points associated with modern theory.

Since the issue of producibility involves the notions of manufacturing confidence and risk, it is reasonable to assert that the concepts underlying probability should serve as the foundation of any measurement scheme. To illustrate why this is so, let us consider an analogous example. For the moment, we shall hypothesize a pair of "manufacturing dice" and a customer requirement which only allows those combinations which yield a 3,4,5...,or 11. In this instance, a 2 or 12 represents a nonconformance to standard, or as some would say, a quality defect. Thus, we may ask the question: "To what extent will the customer be satisfied?" In the language of statistics, the question would be: "What is the probability of not rolling a 2 or 12?" In order to answer the latter question, we must apply some fundamental probability theory.

As many of us may recall, the likelihood of some event A is given by P(A). If some event A is independent of some other event, say B, the probability of the A and B occurring is P(A) X P(B). In other words, the joint probability of A and B is multiplicative by nature [Note #3]. Since a single die has six sides, the random chance probability that any given side will be face up is 1/6 =.1667 because; a) only one side can be up at any given time and b) there are a total of six possibilities. In this case, the random chance probability that any particular side will be face up is 16.67%. Given that we have a pair of dice, we may say now that the probability of rolling two 1's would be .1667 X .1667-.0278, or 2.78%. The same reasoning would also hold true for the possibility of two 6's. This may be directly verified by studying the exhaustive combination given by a pair of dice, such as displayed in figure 4.

Because the occurrence of a 2 or 12 can not happen concurrently, we may say that the two outcomes are mutually exclusive of each other. This is to say that a 2 and 12 can not occur at the same time; i.e., they are restricted from occurring together [Note # 4]. Hence, the events of concern are mutually exclusive. Any time that two events, say A and B are mutually exclusive, the probability of event A or B occurring may be given by summing their individual probabilities [Note # 5]; e.g., P(A)+P(B). Therefore, the total probability of not meeting the customer's standards, with respect

PROBABILITY OF TWO 1s = 1/6 X 1/6 = 1/36 = 0.028
+ PROBABILITY OF TWO 6s = 1/6 X 1/6 = 1/36 = 0.028
PROBABILITY OF NONCONFORMANCE = 2/36 = 0.056

DIE 2 \ DIE 1	1	2	3	4	5	6
1	2	3	4	5	6	7
2	3	4	5	6	7	8
3	4	5	6	7	8	9
4	5	6	7	8	9	10
5	6	7	8	9	10	11
6	7	8	9	10	11	12

Figure 4. Possible Combinations given by a Pair of Dice

to our manufacturing example, would be .0278 +.0278 =.0556, or 5.56%. This represents the risk of nonconformance. Intuitively, we may reason that the likelihood of yielding a 3,4,5,...,or 11 may be given by 1 -.0556 =.9444. In the context of our example, this would be to say that the likelihood of customer satisfaction is 94.44%. Additional insight into these concepts may be realized by examining figure 5, using figure 4 as a frame of reference.

Figure 5. Bar chart of the Probabilities associated with a Pair of Dice

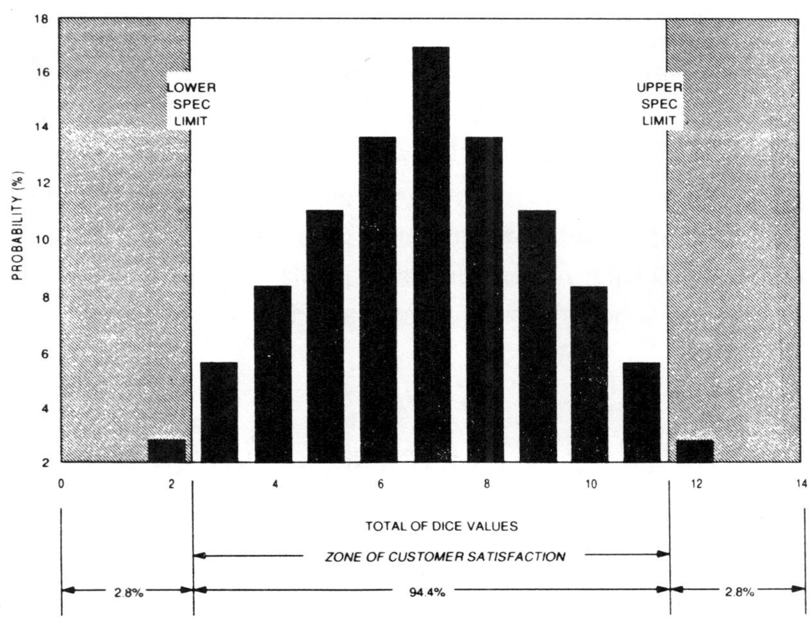

Extending the Producibility Game

Now let us suppose that we have two sets of dice. Let us also hypothesize that the first set of dice is the gaming vehicle for the previously mentioned specification (e.g., the customer has defined a 2 or 12 as a defect). As already indicated, the yield related to the first specification was determined to be 94.44%. We shall now say that the second set of dice is related to a different specification. In this instance, the second specification is given by the combinatorial possibilities yielding a 4,5,6,..., or 10: i.e., a 2,3,11, or 12 would be considered defective. By the same statistical reasoning process we could easily determine that the expected yield, as it applies to the second specification, would be 83.33%.

At this point, we could ask the question: "What is the probability of simultaneously meeting both specifications?" As you will recall, when two events are independent, the joint probability is multiplicative by nature. In this instance, the likelihood of conformance to both specifications can be described by [1-P(A)] x {1-P(B)]. As applied to our example situation, the probability of picking up both sets of dice and rolling a "good" number would be 0.9444 x 0.8333=0.7870. This would be to say that there is a 78.70% chance of simultaneously producing numbers which conform to both specifications. In short, the likelihood of success (i.e., first-time yield) is a direct measure of producibility; if the joint, or "rolled," yield is high, producibility is high; otherwise, producibility is low. The logic of this is quite intuitive.

Probability and Product Parameters

We may now generalize the concepts underlying our gaming example to an actual product parameter. To do so, let us suppose that we have before us some bilateral product performance specification related to parameter Y_1. Here again, we can let Y_1 be whatever comes to mind; e.g., the length of a mechanical part, voltage in a circuit, resistance of a material, etc. Let us also postulate that the underlying distribution is normal [Note # 6]. If we say that event A represents a nonconformance to standard and the probability of encountering A is 0.27%, then the "first-time" yield of parameter Y_1 would be 1.0000-.0027=.9973, or 99.73%. [Note # 7]

If the yield of some other independent characteristic, say, Y_2 was 0.8845, we may argue that the joint probability of conformance to specification, with respect to Y_1 and Y_2 would be 0.9973 X 0.8845-0.8821, or simply 0 88.2%. Notice that the joint yield is also called "rolled-throughput yield [Note # 8]," and designated as Y_{RT}. It is referred to as such because the first-time yield values are mathematically rolled into a single value, where that single value is indicative of the likelihood of non-incidental throughput. Let the reader understand that the terminology does not infer that manufacturing is one big crap shoot, although some may adamantly disagree.

THE FUNDAMENTAL METRICS FOR DESCRIBING PRODUCIBILITY

Estimating Long Term Producibility

Given the review of basic probability theory and its application to producibility, we may now immerse ourselves in an example situation which describes the basic metrics and the methodology of implementation.

For purposes of discussion, let us consider the example product illustrated in figure 6. In this case, the product is a military survival radio. We can readily imagine that this product consists of several major assemblies, with each individual assembly consisting of several subassemblies. Specifically, our discussion will focus on only one of the assemblies displayed in drawing 4 of figure

6. In particular, we will consider one of the related subassemblies as well as the printed circuit board (PCB) illustrated in drawing 3 of figure 6. As may be apparent, only one chain in the total design hierarchy will be used to illustrate and support the ensuing discussion. This has been done to delimit the scope of presentation. With little study, one can readily generalize the ensuing logic to any other chain in the design hierarchy, or any other product for that matter.

With respect to the subassembly, it will be known that there are i mechanical specifications; e.g., dimensional tolerances, stress requirements, thermal limits, etc. In addition, we shall say that there are j electrical specifications; e.g., response time, circuit gain, etc. We shall also say that there are k other characteristics which must conform to specification; e.g., quality characteristics such as solder joint wetting, paint finish, etc. Thus, we may now say that there are i+j number of test parameters and k number of inspection characteristics. The reader should recognize that the latter classification is arbitrary; however, in the context of this particular product example, the categories will take on applied meaning as the discussion unfolds.

The Role of Parameter Capability Data

Now, imagine that we have before us a database of process, material, and component capability data. [Note # 9] We shall also note that the database was formulated on the basis of extensive characterization studies [Note # 10]. Let us further postulate that the aforementioned data is of two varieties; namely, short-term and long-term [Note #.11] Next, we shall say that appropriate computer analyses and simulations were undertaken for each of the i and j parameters using the characterization data as a guide. Obviously, such an undertaking would produce more realistic short- and long-term first-time parameter yield estimates simply because we would be working with the actual process, component, and material capabilities.

The Role of Mechanical Engineering

In order to better understand the implications of the latter discussion, we shall return to our radio example. Suppose that we select one of the assembly gaps (gap 1) as related to, say, subassembly K. If the process capabilities of the related parts are known, or capable of rational estimation, it would be possible to compute the probability of assembly. [Note # 12] This would be done using a statistically based methodology such as described by Harry and Stewart (1988). [Note # 13] Using this type of analytical procedure would reveal the expected first-time yield of the gap. Recognize that this could be done for various other subassemblies.

If there were a total of x=6 independent assembly gaps related to subassembly K, and the long-term first-time yield(Y_{FT}) projection for each gap was 99.73%, it is easy to see that the rolled-throughput yield (Y_{RT}) for the product subassembly, over many manufacturing intervals, would be $Y_{RT} = Y_{FT}{}^{X}.9973^{6} =.9839$, or 0 98.39%. [Note # 14] This is to say that there would be a 98.39% likelihood that any given product subassembly could be put together without encountering an interference fit.

As we can now see, it is possible to estimate the joint yield or "simultaneous probability" of zero irregularities with respect to all j mechanical specifications. [Note # 15] As may be apparent, our example excludes all other mechanical considerations with the exception of assembly performance; i.e., i=x. This has been done in the interests of simplicity and reading ease. Let the reader recognize that the inclusion of other yields, such as those related to thermal and stress requirements,

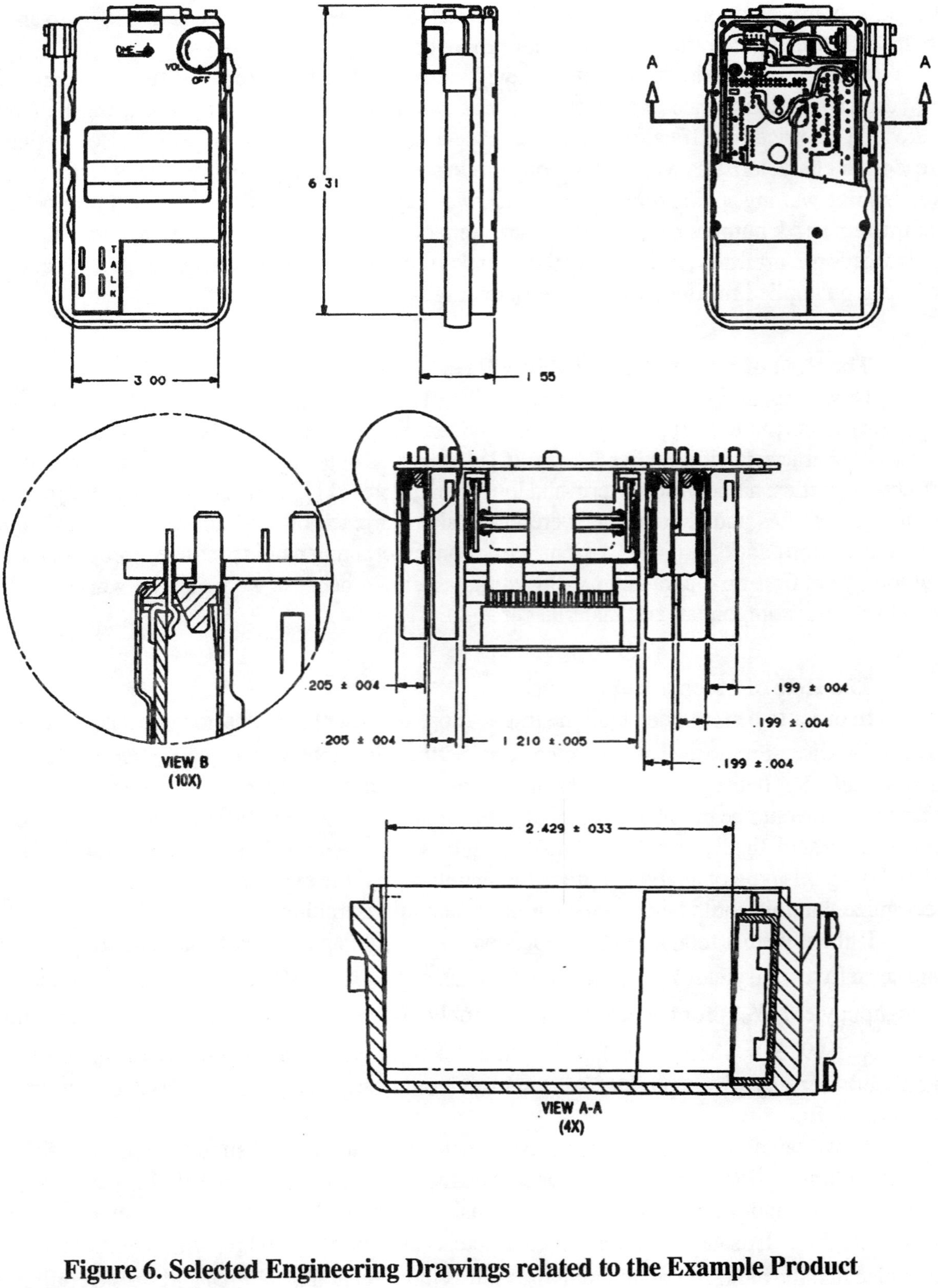

Figure 6. Selected Engineering Drawings related to the Example Product

or any others for that matter could easily be incorporated into the analysis via the same methodology. In turn, an overall rolled-throughout yield could be computed so as to establish the overall likelihood of success, with respect to the mechanical requirements. For purposes of furthering our discussion, we shall assume that each of the product subassemblies contains i=x=6 assembly gap specifications.

The Role of Electrical Engineering

In the context of electrical circuit performance, let us turn our attention to PCB illustrated within drawing 3 of figure 6. In this case, we shall say that this particular PCB consists of several circuits. Let us consider just one of those circuits, say circuit V. Suppose that V was intended to output a certain electrical response, say voltage. Let us also assume that it is possible to simulate the response (e.g., voltage) using a computer simulation program.

If we were to input the known (or rationally postulated) statistical characteristics of each part (related to circuit V) into the simulation, it would be possible to study the effect of component variation on voltage. From such study, the projected first-time test yield of circuit voltage could be ascertained. Again, if there were a total of j=6 independent electrical response parameters associated with the PCB, and the long-term first-time yield (Y_{FT}) projection for each of the six parameters was 99.73%, the rolled-throughout yield (Y_{RT})for the PCB, over many manufacturing intervals, would be $Y_{RT} = Y_{FT^J} = .9973^6 = .9839$, or 98.39%. This is to say that there would be a 98.39% likelihood that any PCB would pass the required electrical tests.

Notice that all of the independent circuit response yields could be treated in the same manner, thereby deriving the rolled-throughput yield for all j electrical parameters. However, for the sake of simplicity and reading ease, we shall assume there is only one PCB per product subassembly and j=6 circuit responses per PCB.

The Role of Manufacturing and Quality Engineering

Now, let us suppose that the subassembly is subject to k inspection characteristics. If the historical defects-per-unit (dpu) data is available within the aforementioned database, or can be rationally postulated for each of the k characteristics, we could estimate the rolled-throughput inspection yield by virtue of the Poisson relation [Note #17]. For example, let us say that the historical soldering performance (as related to solder joint wetting) was such that dpu=.000006. Notice that, in this instance, a unit is defined as an individual solder joint. Given this, the expected first-time wetting yield would be $Y_{RT} = Y_{FT^K} = .999999016^6 = .9973$, or 99.73%. If the same rolled-throughput yield applied to each of the k=6 inspection categories, we would readily determine that the long-term probability of zero defects, as related to any given subassembly, would be $Y_{RT} = Y_{FT^K} = .9973^6 = .9839$, or 98.39%. This would hold true since there is (by definition) only one PCB per subassembly.

Estimating the Composite Long-Term Design Producibility

The overall long-term first-time yield for all i,j, and k product subassembly parameters could be given by 0.9839 X 0.9839 X 0.9839 = 0.9525. This is to say that here would be 95.25% confidence of "specification conformance" over a great many production intervals as related to the electrical, mechanical, and inspection aspects of the product subassembly.

If the same estimate applied to each of the t=6 subassemblies, the product rolled-throughout yield would then be $Y_{RT} = Y_{FT^t} = .9525^6 = .7467$, or 74.67%. Hence, we now have a relative benchmark to gage the producibility of the composite product design. The reader should also be aware that manufacturing cycle time, as well as other manufacturing and business issues, can be studied at this point in the analysis [Note #18].

Comparison of Design Alternatives and the Benchmarking Process

Let us now turn our focus toward what is called "normalized rolled-throughput" yield ($Y_{RT;N}$). Essentially, such a yield represents the throughput-per-parameter, or part, process step, etc. In this sense, it may be thought of as the first-time yield of a "typical" parameter. Essentially, normalization has the effect of neutralizing the influence of complexity. Because of this feature, it is possible to compare the producibility of one design alternative to another. In short, it gives us the ability to compare "apples and oranges," so to speak.

To illustrate how this works, let us extend the previous example. Since the long-term product level rolled-throughput yield for this particular design (alternative A) is 0.7467, and given there are m=5000 parts, then the normalized long-term rolled-throughput yield per part would be ${Y_{RT;N}}^{1/m} = .7467^{1/5000} = .999942$, or 99.9942%. This is to say that there would be 99.9942% confidence that any given part could go through the entire manufacturing process without a quality related incident. The interactive effect of manufacturing capability and product (or process) complexity on rolled-throughput yield can be better understood by the graphs presented in figures 7a and 7b.

The reader should notice that the capabilities given in figure 7b have been sensitized so as to account for the effects of long-term perturbations in process centering. For more information on the "adjustment factor," the reader is guided to Harry and Prins (1991).

If we were to equilibrate the latter yield value to cumulative area under the standard normal curve, we would discover that 0.999942 area is approximately equivalent to 3.85σ for the unilateral case [Note # 19]. Thus, the long-term producibility benchmark (per part) for design alternative A would be 3.85σ. Remember that the long-term goal for Motorola Inc. is 4.5σ. As may be apparent, such "sigma" performance reduces to $C_p = 3.85 / 3.00 = 1.28$.

Now, let us turn our attention to design alternative B. If design B revealed a long-term rolled-throughput yield estimate of .98 and consisted of m=4000 parts, the normalized long-term rolled-throughput yield per part would be ${Y_{RT;N}}^{1/m} = .98^{1/4000} = .999995$, or 99.9995%. This would be equivalent to approximately 4.42σ. In terms of capability ratio, $C_p = 4.42 / 3.00 = 1.47$. Obviously, design B is superior to design A in terms of expected long-term producibility and, consequently, would be considered more as a viable alternative over many intervals of manufacturing.

Estimating Short-term Producibility

We shall now consider the producibility analysis of design alternative A under the constraint of short-term capability data (as related to the process, material, and components). Essentially, whenever a producibility study is based on this type of parameter capability data, the analytical output reflect "instantaneous producibility." In other words, it describes the capability of any given

Figure 7a. The Unilateral effect of Short-Term Capability and Complexity (X axis) on Rolled-Throughput Yield (Y axis)

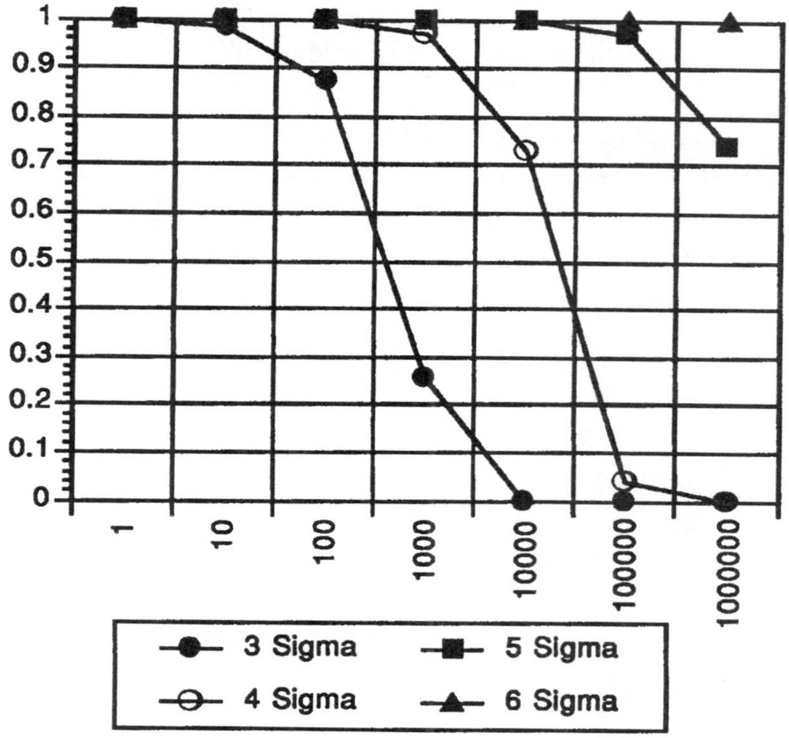

Figure 7b. The Unilateral effect of Long-Term Capability and Complexity (X axis) on Rolled-Throughput Yield (Y axis)

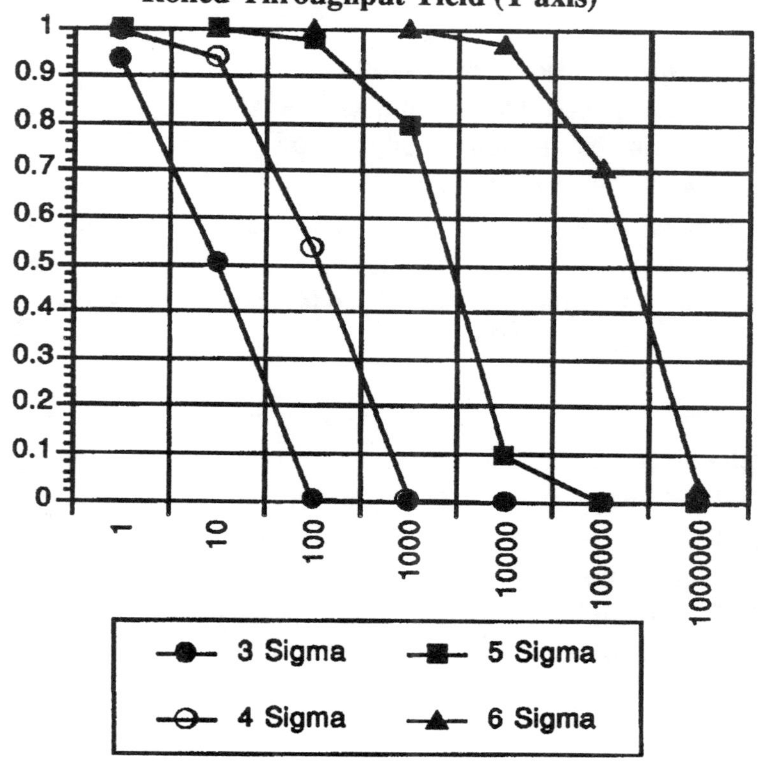

parameter free of perturbing temporal influences. Hence, if the short-term capability data is used as the basis of computer modeling, the outcome projects what would be expected when only random variation is present.

To illustrate, lets postulate that the widget simulation was conducted using the short-term capability database associated with the i,j, and k performance parameters. In this context, let us further suppose that Y_{RT} for i,j, and k was 0.99994, 0.99998, and 0.99962, respectively. From this, we may conclude that the joint yield would be 0.99962 X 0.99998 X 0.99962 = 0.99954, or 99.95%. Again, we may find the normalized rolled-throughput yield by virtue of $.99954^{1/m}$, where m=5000. In this instance, the short-term rolled-throughput yield expectation would be $Y_{RT;N} = .99954^{1/5000} = .9999999$, or approximately 99.99999%. Calibrating this to area under the standard normal curve would reveal that .9999999 cumulative area is given by 5.21σ. Hence, the short-term or "instantaneous" producibility benchmark is 5.21σ.

Analysis of Design Robustness

Now, if one were to consider the discrepancy between the normalized short-term and long-term producibility estimated, the degree of design "robustness" could be readily ascertained. To exemplify, let us further study design alternative A. Since the short-term estimate of producibility was 5.21σ (per part) and the long-term was 3.85σ (per part), it may then be reasoned that the discrepancy was $5.21\sigma - 3.85\sigma = 1.49\sigma$.

We may now say that 1.49σ represents an equivalent mean shift. The reader is directed to Harry and Prins (1991) for a thorough discussion of the mathematical details pertaining to the equivalent mean shift. In this instance, the design robustness may be given as $(3.85\sigma / 5.21\sigma)100 = 73.9\%$. This would be to say that the design is 74% robust to perturbing influences of a temporal nature. On the flip side, we may say that the instantaneous design producibility was degraded by $(1.00\text{-}.74)100 = 26\%$.

Producibility Optimization Alternatives

It is easy to see how m could be the number of process steps, operations, specifications, etc. In fact, m could just about be anything related to complexity. In essence, normalized rolled-throughput yield allows us to study producibility from many different perspectives. However, only such traditional variables as part count and process operations could be used as the basis of normalization. Also note that m can be weighted by relative cost: however, it must be recognized that most weighing schemes introduce an additional layer of computational complexity. This is not to say that cost should be disregarded, but rather to say that there are better ways of studying it [Note # 20].

If any given design fails to reveal a satisfactory index of producibility, several alternatives are available. First, one could elect to hierarchically study the producibility metrics; i.e., create a Pareto breakdown of the design so as to isolate the leverage. Once the leverage is isolated, one can "what if" the baseline design via the manipulation of part count and manufacturing capability. For example, one could explore possible product and/or process design changes that would result in a reduced complexity factor; e.g., m could be minimized such that the desired rolled-throughput yield was obtained. Also, various aspects of the design could be optimized in terms of robustness. Another exploratory procedure would be to artificially minimize the process, component, and material capabilities so as to determine the feasibility of a manufacturing control solution. Yet another alternative would be some combination of all three.

Regardless of optimization strategy, it is quite clear that the independent and joint effects of design complexity (product and process) and manufacturing capability largely determine the functional producibility of a product design. The reader must recognize that "postmortem' optimization efforts can be avoided via a pro-active design process. For example, the methodology presented in this section of the paper can be worked in reverse so as to establish initial, baseline design goals [Note # 21].

SYNOPSIS OF THE PRODUCIBILITY ASSESSMENT MODEL

As a consequence of the aforementioned arguments, we can now see how rolled-throughput yield and robustness can be related to all three of the circles given in figure 3. However, it is also quite apparent that such a system of analysis is highly dependent upon timely and accurate capability data. But how do we go about getting such data? And what indices should the data be reduced to? In order to answer such questions, we must delve into the issues of process, component, and material characterization.

As a final point on the topic of producibility metrics, the reader is again reminded that any given quantitative assessment of producibility is only as good as the data and assumptions which underlie the final numbers. The age old saying "garbage in, garbage out" certainly applies to the methodologies proposed in this paper. In addition, we are all reminded of the precarious balance which must exist between the real world of practical constraints and the abstract domain of the mathematical precision. Too much of the former can easily invalidate a quantitative analysis. On the other hand, too much of the latter can lead to perpetual "analysis paralysis."

THE CHARACTERIZATION OF PROCESSES, COMPONENTS, AND MATERIAL

The Overall Perspective

Essentially, the characterization strategy consists of 4 primary phases. Specifically, those phases are: 1) product parameter definition, 2) product parameter analyses, 3) process parameter optimization, and 4) process parameter control. As may be apparent, this strategy seeks to link the process to the product in a scientific manner. The basic structure of the four phase approach has been located in figure 9.

Phase I of the Strategy

In overview of the characterization strategy, we may say that the first phase involves defining the critical product response characteristic to be studied. Although more than one response variable may be identified and subsequently studied, we shall constrain our discussion to only one. Again, this has been done for the sake of reading ease and simplicity of communication.

Figure 9. Process Characterization Strategy

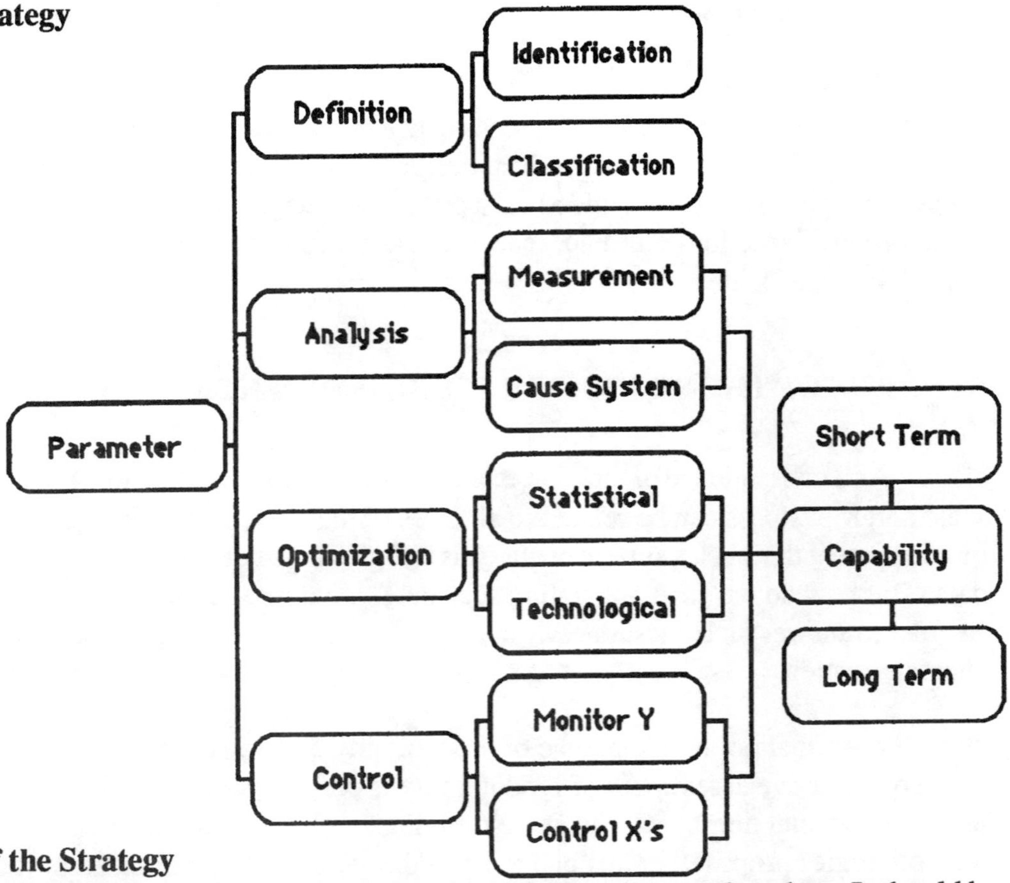

Phase II of the Strategy

The second phase of the strategy entails the collection of baseline product data. It should be pointed out that the baseline data is gathered on a short and long-term basis. The "short-term" data is used to estimate the "instantaneous reproducibility" of the response characteristic (Y). In this manner, it is possible to establish the extent to which the process is capable of meeting the specified product quality standards—free of perturbing influences.

The "long-term" data is used to estimate the "sustained reproducibility: of the dependent variable over a very large number of production intervals. Intuitively, one may reason that such data not only reflects instantaneous or "random" noise, but systematic, nonrandom influences as well. From these two measures, it is possible to study the extent to which Y is robust to perturbing influences within the related network of causation. From another vantage point, we may view how well the process is controlled over time.

Once the product parameter capability has been established, the second phase is continued by identifying the key process variables (X1...Xn) within the network of causation which exert an undue influence on Y. This is most often accomplished through the application of statistically designed experiments and various diagnostic tools. The intent of such activity is three fold; 1) identify leverage with the network of causation, 2) study competitive settings among the major causal variables, and 3) assess the feasibility and relative efficacy of control with respect to the key casual variable. Of course, the capability may need to be reassessed after such activity.

Phase III of the Strategy

The third phase involves optimizing the product response characteristic in relation to the given product standards and specifications. This particular phase would be exercised only if the parameter capability data and manufacturing situation so warranted. The intent is to "fine tune" the independent variables identified during the second phase. In addition, this phase focuses on desensitizing the response characteristic to systematic and non-systematic sources of variation. Because of this, the application of advanced inferential statistics, response surface design, and computer modeling is often required. Here again, it may be desirable to reassess capability.

Phase IV of the Strategy

Finally, the fourth phase is related to monitoring the dependent variable (Y) and controlling the "vital few" process variables (X's) within their respective operating tolerances. After a "settling in" period, the product and process parameter capabilities would be reassessed. Depending upon the outcomes of such a follow-on analysis, it may be necessary to revisit one or more of the preceding phases.

As we can now see, the process characterization strategy is a scientific and deductive method for structuring a product response problem. As may be noted, the strategy emphasizes deductive inquiry and repeatability, two of the central issues associated with scientific investigation. In this sense, the strategy constitutes "working smarter, not harder."

When this process is applied across many different products, an organization possesses the ability to benchmark its overall design and manufacturing ability. Through the careful analysis of such benchmarking information, the organization can discover how to capitalize on its strengths and overcome areas of limitation.

NOTES

[1] At all times, the reader must remain cognizant of the fact that the field of producibility assessment is relatively new. Because of this and its enormous scope, narration and illustration of many details is not possible within the confines of this paper. As a consequence, emphasis is placed on the development of a conceptual understanding at many points in the discussion.

[2] At a more global level, the reader should recognize that Motorola, Inc. considers the use of 'metrics' to be the foundation of its success. Within Motorola it is well recognized that the periodic review of key business indices or 'metrics' causes people to focus on these things which are essential to the business. Obviously, if people are not focused, improvement will be significantly more difficult. Some of the metrics are inclusive of, but not limited to: cycle time, field performance, delivery delinquencies, etc. With such data, it is possible to institutionalize a standardized feedback control system. For more detailed description of this process and related metrics, the reader is

directed to the GEG Performance Metrics Manual published by the Government Electronics Group, Motorola, Inc. Scottsdale, AZ.

[3] This is given by the multiplication theorem. This theorem states that the probability of several particular outcomes occurring jointly is the product of their separate probabilities, provided that the events that generate these outcomes are independent. If they are not independent, the joint occurrence of both A and B would be given by P(A and B) = P(A) x P(A/B), where P(A/B) is equal to the probability that B will occur assuming A has already occurred.

[4] Such would be the case when considering a bilateral specification. A given measurement can not be greater than the upper specification limit (USL) and less than the lower specification limit (LSL) at the same time. This observation is nonconforming to the USL or LSL, but not both.

[5] This is given by the addition theorem. This theorem states that the probability of occurrence of any one of several particular outcomes is the sum of their probabilities, provided they are mutually exclusive. If the events are not mutually exclusive, the probability of A or B occurring would be given by P(A or B) = P(A) + P(B) - P(AB), where P(AB) is the joint probability of A and B; e.g., P(AB) = P(A) x P(B).

[6] The uninformed reader must recognize that a comprehensive discussion of the normal density function, or any other distribution for that matter, is far beyond the scope and intent of this paper. Let it suffice to say, that the normal distribution is omnipresent with respect to many industrial applications of mathematical statistics. It is characterized by its symmetrical bell shape and, for this reason, is often called the "bell curve." Regardless of labels, it characterizes the frequency of occurrence of a parameter relative to its measurement scale. As a general rule (where producibility analysis is concerned) the normal distribution should be applied unless direct evidence or sound reasoning indicates otherwise. The reader is directed to the bibliography should additional information on this topic be required.

[7] The example assumes an underlying distribution which is normal and independent. This is often a reasonable assumption; however, if it is not, mathematical transformations may be utilized so as to artificially force the assumption. While there are other ways to mathematically treat such problems, it is beyond the scope and intent of this paper to articulate those methods. The reader is directed to the bibliography should additional information on this topic be required.

[8] Rolled-throughput yield is the product of all subordinate yields. For example, if there were three independent mechanical parameters with respective yields of .999, .998, .990, the rolled throughput yield would be .999 x .990 = .9870. The rolled-throughput yield represents the likelihood of zero nonconformities. In essence, it is the cumulative yield for k number of independent yield values. Of course, many would argue that some characteristics are not independent of others and, as a consequence, the cumulative yield would be biased. While this is a true statement, we must recognize the balance between practical application and mathematical precision. In many instances, the covariance of two (or even three) events can easily be accounted for; however, if the number of covariates is large due to the total number of events being considered, the task of computation can quickly overwhelm most practitioners. If one is concerned with "seven digit precision," then the task

of quantitative producibility analysis will be prolonged, and perhaps even prove to be a frustrating journey. Even if such precision is attainable, it is doubtful, in most instances, that the added value of precision in the final figures will justify the lower order concerns. While this certainly does not hold in all cases, it does hold in many. Therefore, as a general rule (where producibility analysis is concerned), the assumption of independence is reasonable unless direct knowledge indicates otherwise.

[9] In the form of statistical information; e.g., mean, variance, equivalent first time yield, parts-per-million (ppm), defects-per-unit (dpu), etc.

[10] Such studies are conducted so as to derive statistical data related to variation. In turn, the statistical data is used to project first time yield. Although the characterization process is overviewed in this paper, the reader must recognize there is "more to it than meets the eye." For example, only certain types of statistical information are useful in quantitative producibility analyses. If an engineer has the "wrong" type of characterization data, it is possible that the engineer could be precluded from conducting a meaningful quantitative producibility analysis. In short, the characterization data serves as the base of the producibility metrics given in this paper. Therefore, we may say that producibility analysis and parameter characterization forms and interactive sub-system within the "design for manufacturability" strategy. Unless "data alignment" is at hand, the likelihood of thwarted analysis is high.

[11] Short-term capability data reflects "instantaneous reproducibility;" i.e., the ability of a process to repeat itself in a short period of time. Note that short-term capability data must only reflect random variation. If so, the resultant variance describes random error. If the data reflects only random error, it is reasonable to say that it is free of perturbing, systematic influences which emanate from the network of causal variables. Such influences are reflected in the long-term capability data. The relative difference between short and long-term indices of capability reflects the extent to which existing process, material, and component controls are effective. In short, it is a measure of how robust the response characteristic is to systematic, nonrandom variations in the underlying system of causal variables; i.e., it is an indirect measure of manufacturing control. From this vantage point, it is easy to see why such data is vital to a quantitative producibility assessment. It is entirely possible that "instantaneous producibility" is high, but low over many manufacturing intervals. Such results can have profound business implications if high volume production is anticipated.

[12] Rational estimation is usually based on extensive engineering and manufacturing experience. For example, it is well known that many manufacturing processes operating under the auspices of SPC display a capability ratios between $1.00 \leq Cp \leq 1.33$; therefore, the pessimistic assumption would be $Cp = 1.00$ and the optimistic assumption would be $Cp = 1.33$. Such assumptions are often (not always) considered "rational" because many of the textbooks on SPC provide this range as a target for purposes of manufacturing. As a consequence, the novice practitioner controls the process to this level, simply because "the book said so." In fact, they are often rewarded for such performance, which does nothing more than reinforce the behavior, not to mention the antiquated textbook — and the cycle goes on and on. As a side line to this discussion, the reader must recognize that many practitioners no longer accept the aforementioned capability range as satisfactory. Today, some organizations require $1.5 \leq Cp \leq 2.0$, or even $Cp > 2.0$.

[13] Specifically, the probability would be ascertained from the maximum and minimum standard normal deviates (Z values) associated with the assembly specifications (i.e., the lower and upper assembly gap specifications). The Z value is found by calculating the ratio of nominal assembly gap (numerator term) and assembly gap standard deviation (denominator term). Notice that the assembly gap standard deviation is derived by the root-sum-squares (RSS) method. The reader is directed to the bibliography should additional information on this topic be required.

[14] If x were large enough to represent a practical constraints, sampling methods could be invoked; e.g,. random sampling procedures. Assuming a random selection process, it would be reasonable to assert that the sample average would be representative of the universe average. Under such conditions, the average first time yield could be used as a base for the rolled-throughput yield estimate.

[15] By direct calculation in the instance of our assembly gap example, or via simulations driven by mathematical algorithms (e.g., for those mechanical characteristics which cannot be conveniently calculated such as is often the case with solids modeling, thermodynamics, etc.). Regardless of the form estimation, the reader must recognize that the output is only as good as the input. By using actual short- and long-term input as the basis of simulation, the outcome is more "intelligent" and, as a consequence, provides a better vehicle for conducting quantitative producibility analyses.

[17] If the probability of a defect is less than 10% but the overall likelihood of observing is high, the Poisson relation will apply in many cases; otherwise, the binomial (or some other distribution) should be considered. Given that the defects are independent and randomly distributed, it can readily be demonstrated that the Poisson relation reduces to $e^{-(d/u)}$. The resultant represents the probability of zero defects; i.e., the expected first time inspection yield for a unit of product with d defects. The reader is directed to the bibliography should additional information on this topic be required.

[18] The impact of product quality on manufacturing cycle time can be expressed as $1+(1-Y_{FT})$, where Y_{FT} is the expected first time yield. This relation gives the additional number of units which must be processed to compensate for not having 100% first time yields. Hence, if the theoretical optimum cycle time can be computed or rationally estimated, it is possible to project a "quality adjusted" cycle time. In turn, this figure can be used to study the impact of quality on such things as inventory costs, equipment capacity, manpower requirements, return on net assists (RONA), etc. It is a very powerful relationship if used correctly.

[19] For additional information on the benchmarking process and analytical conventions, the reader is guided to Harry and Lawson (1988).

[20] For example, it is possible to apply the quadratic loss function to illustrate the unit change in economic loss as a function of some unit in parameter variation, complexity, or both. The reader is directed to the bibliography should additional information on this topic be required.

[21] For example, one can establish a desired rolled-throughput yield for all of the "test" parameters and then back compute the maximum allowable part count, given a per part capability figure, via an nth root transform. In the instance of process design, the inspection rolled-throughput yield target can be decomposed and subsequently allocated in the same fashion. When this is done, producibility can be structured prior to the creation of engineering drawings; i.e., before the design gets "locked in concrete."

REFERENCES

Bender, A. (1975). "Statistical Tolerancing as it Relates to Quality Control and the Designer." Automotive Division Newsletter of ASQC.

DoD(1989), DoD 500.51-G. Total Quality Management: A Guide for Implementation, March 15, 1989.

Evans, David H., (1974). "Statistical Tolerancing: The State of the Art, Part I: Background," Journal of Quality and Technology, 6(4), pp. 188-195.

Evans, David H., (1975). "Statistical Tolerancing: The State of the Art, Part II: Methods for Estimating Moments," Journal of Quality and Technology, 7(1), pp.1-12.

Evans, David H., (1975). "Statistical Tolerancing: The State of the Art, Part III: Shifts and Drifts," Journal of Quality and Technology, 7(2), pp.72-76.

Gilson, J., (1951). "A New Approach to Engineering Tolerances," Machinery Publishing Co., Ltd., London.

Grant, E. L., and Leavenworth, R. S. (1972). Statistical Quality Control (4th Edition). New York: McGraw-Hill Book Company.

Harry, M. J. (1986). "The Nature of Six Sigma Quality". Motorola University Press, Motorola Inc., Schaumberg, Illinois.

Harry, M. J. and Lawson, R.J. (1988). "Six Sigma Producibility Analysis and Process Characterization." Publication Number 6Sigma-3-03/88. Motorola University Press, Motorola Inc., Schaumberg, Illinois.

Harry, M. J. and Prins, J. (1991). "The Vision of Six Sigma: Mathematical Constructs Related to Process Centering." Publication Pending. Motorola University Press, Motorola Inc., Schaumberg Illinois.

Harry, M. J. and Stewart, R. (1988). "Six Sigma Mechanical Design Tolerancing." Publication Number 6Sigma-2-10/88. Motorola University Press, Motorola Inc., Schaumberg Illinois.

Hillkirk, J. (1989), "Top Quality is Behind Comeback", Cover story, USA Today, March 28, 1989.

Juran, J. M., Gryna, F. M., and Bingham, R.S. (1979), Quality Control Handbook, New York, NY: McGraw-Hill Book Co.

Kelvin, W. T., 1891, Popular Lectures and Addresses.

Motorola Inc. (1986), "Design for Manufacturability": Eng 123 (Participant Guide), Motorola Training and Education Center, Motorola Inc., Schaumberg, IL.

Mood, A. and Graybill, F. (1963), Introduction To The Theory of Statistics (2nd Edition). New York: McGraw-Hill Book Co.

Pearson, E. S. and Hartley, H.O. (1972), Biometrika Tables for Statisticians, Vol. 2, Cambridge University Press, Cambridge.

Shewhart, W. A. (1931), Economic Control of Quality of Manufactured Products, Van Nostrand

Company, Inc.
Stoll, H., Kumar, A., and Maas, D., (date unknown), " Producibility Measurement: A key to the Design of Producible Products", Unpublished paper, pp.1-2, Industrial Technology Institute, Ann Arbor, MI.
U. S. Navy, (1989), Producibility Measurement Conference, Jan. 24-25, 1989, Scottsdale, AZ.

APPLICATION OF TAGUCHI ROBUST DESIGN METHOD TO THE DESIGN OF FRICTION RETARD FEEDER

Louis LaVallee*

SUMMARY

Taguchi's Quality Engineering methods (also called Robust Design methods) differ from the traditional design engineering methods. The traditional engineering approach utilizes 'design-build-test-fix' sequence to design products consisting of a number of subsystems. In this approach, one design concept is selected based on expert opinions and a series of prototypes are built, tested, and fixed until it is time to start production. The robust design methods, on the other hand, focus on making sure that the design parameters are chosen to minimize the influence of the variabilities of user environment and manufacturing process on the main function of the design. Also, the tolerance design in the robust design methodology is done to minimize the added cost of tightening the tolerances only after improving the main function through parameter design. Upon improving the robustness of the subsystems through orthogonal series of tests, only one or two prototypes are necessary prior to starting production. This greatly reduces the time for new product introduction.

A comparison of these two design engineering approaches is presented with an example from a recent Xerox 5034 paper feeder design activity. Some of Taguchi's ideas on quality loss, what to measure, and how to improve product robustness are presented. The operating window methodology, as described by Taguchi and Clausing, is shown to be a powerful tool for early prototype hardware. The best nominal values for paper feed design factors were identified from parameter design. Experimental efficiency and an early focus on design robustness enables shorter product delivery time, improved reliability, and higher customer satisfaction.

* The author is with Xerox Corporation, APT & E/ M &OT /Design Methods & Standards, Webster, NY.

INTRODUCTION

Xerox Corporation has been one of the prominent users of Genichi Taguchi's quality engineering methods in U.S. industry. Through Xerox's direct involvement with Fuji Xerox in the early 1980's, the reputation of Taguchi's work became known. Fuji Xerox's ability to deliver low cost quality products to market in one half the time of Xerox, was strong incentive to do internal benchmarking. What was found was that Taguchi had developed a comprehensive systematic methodology for improving engineering design process, not so much for problem solving, but for problem prevention. Fuji Xerox had integrated the Taguchi methodology into a "Phased Program Management" product delivery process (developed by NASA during 1970's), which in a simplified form, is still in use at Fuji Xerox company.

Genichi Taguchi was brought to Xerox Corporation by Don Clausing in 1982 for consultation and initial training of twenty engineers, including the author. Shortly after his initial visit, the Taguchi parameter design methods were applied to several pilot manufacturing processes which had been having serious yield problems. The results were quite remarkable. Wayland Hicks, then the head of Product Development and Manufacturing, met with Taguchi several times and thereafter acted as a strong advocate for deployment of the methodology. In 1985, the Xerox Design Institute was formed. Its mission was to first identify, prioritize, and develop the best design practices for the effective implementation of an engineering design process, and second, to educate the engineering community in the most effective use of the new design practices. Taguchi was appointed to the board of directors of the Design Institute. Biannual visits by Taguchi helped develop more case studies and further understanding of the methodology and philosophy. Each year there was a steady increase in the number of applications. Today, many hundreds of applications are performed annually.

TRADITIONAL APPROACH

The traditional engineering approach to designing products consisting of subsystems is the design-build-test-fix scenario, as shown in figure 1, where one design concept is selected (based on expert opinion), and a series of prototypes are planned and built, tested, and fixed until they function well at nominal conditions per the design intent interpreted by manufacturing. Early prototype hardware usually has little flexibility to explore large areas of design space. High grade materials and tight tolerances are specified with the thought that down-costing can be done later. Testing of the design usually consists of measuring all dysfunctions, i.e., what the design is not supposed to do. For example, in the case of a paper feeder, this would include:

1. Measurement of paper damage, toner image on paper damage (smudge rating), jam rate (jams/million copies fed), misfeed rate, number of extra sheets in multifeed slugs, other subjective rating schemes, etc., followed by multi-criteria design optimization methods.

2. Determination of the locus of failure boundaries using one factor at a time methods with considerable effort spent on finding the center of mass (the best combination) of the resulting operating window to avoid failures. Avoiding failures and searching for root causes of failures when they do occur, is the main thrust of most traditional engineering development activities.

3. Describing the system analytically using the best available physical and empirical models, collecting lots of data to validate model assumptions, and using some form of simulation like Monte Carlo to reintroduce variability (such as caused by deviations and tolerances of materials and processes) to predict performance.

Traditional engineering practice usually involves one factor-at-a-time testing for gathering information about the design limitations and performance, with all other factors held constant. The objective of many of these tests is to understand underlying mechanisms of deterioration, failure, and dysfunction. Some engineers, (usually with a modicum of statistical training), prefer full factorial design testing with three or four factors at a time (following the well known methodology of Design of Experiments) since interactions among design factors are a major concern. One can generate very attractive data suitable for regression model fitting using this method. Still, other engineers use an ad-hoc approach for design testing, relying on the statistical experts to make some sense out of the data later. The role of the statistician in this case is to perform postmortems on the data collected. Where multiple responses are measured, simultaneous optimization is done, usually with some arbitrary data weighting scheme for handling tradeoffs.

Feedback or feed-forward (sometimes model-based) control systems are included in many subsystems to control functional variability. These add complexity to the designs which may affect cost. Usually, software development also adds cost and delays delivery time. Control systems may ultimately be necessary, but many times improving robustness through parameter design minimizes the need for control systems, or at least reduces the power required to compensate the output variation. The robustness of the sensing systems themselves can also become an issue.

Reliability specifications have been given for most designs, so extensive life testing is being continually done to assure conformance to life goals. For example, paper feeders are run day and night to count the number of misfeeds and multifeeds in every million copies. Decisions to either continue development of the design or to scale up the manufacturing process are based on gut feel with an expectation that the technology will mature later. Extreme conditions of predicted design usage, such as heavy paper and light paper, upcurl and downcurl paper, low stacks and high stacks of paper, are usually deferred to later testing to avoid further funding problems. Preparations for the next demonstration test for management's continued support is the main focus.

Physical, mathematical, or geometric models describing the hardware to be built are sometimes developed in an effort to minimize or eliminate hardware built-test-fix iterations. The models, describing input/output relationships for systems, subsystems, subassemblies and components, are usually quite complex and include many untested assumptions. Much of the subsequent experimental work is aimed at either "characterizing" the effects of the many factors affecting the output. (To characterize 'noise' (i.e., uncontrolled variability) is quite different from making the design insensitive to noise) or validating the many assumptions of the model. Focus on improvement of the hardware and rapid product or process development, many times is lost.

Most models do not differentiate between noise factors and control factors. Classical signal-to-noise ratios, such as the one shown below, are sometimes calculated, but suffer from several problems including:

$$\eta = \frac{(\frac{\partial f}{\partial m})^2}{(\frac{\partial f}{\partial x_1})\sigma_{x_1}^2 + (\frac{\partial f}{\partial x_2})\sigma_{x_2}^2 + \ldots . (\frac{\partial f}{\partial x_n})\sigma_{x_n}^2} = \frac{\text{Signal}}{\text{Noise}}$$

1. Partial derivatives are different at different design space locations
2. Higher order terms are ignored
3. Distortion is not considered

4. Difficult calculation without complete response surface definition
5. Lots of derivatives
6. Monte-Carlo methods for each y (dependent variable) for simultaneous optimization.

Deterministic models generally only predict the mean response. Variation is usually handled by simulation. Monte Carlo methods are quite popular, as they allow any and all factors in the model to be treated as random variables. Probability distribution functions (PDF) for each of the variables are usually assumed, e.g., normal, uniform, log-normal. The independence of the various PDF's is also usually assumed. If the simulation results show no out of specification or defective conditions, the engineers are led to believe that the design is a good one. Verification of simulation results with test data is recommended, but is very difficult to realize in practice.

When specifications are met, the traditional approach is to discontinue further engineering improvement efforts. The design is considered to be ready for scale up when laboratory testing shows "acceptable levels" of defects. Reducing scrap and rework cost is another major objective. However, this is usually left to manufacturing and production.

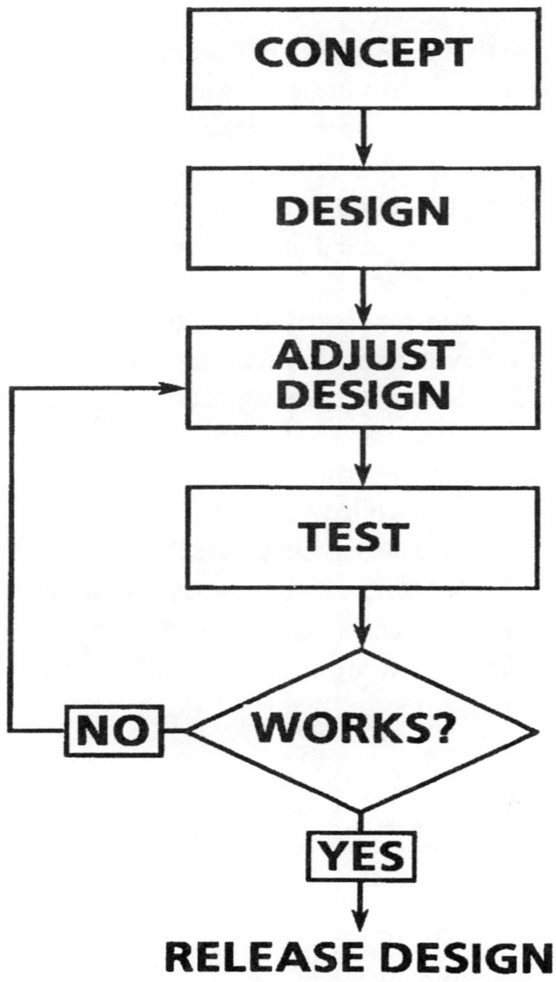

Figure 1. Traditional Design Cycle

TAGUCHI QUALITY ENGINEERING METHODS FOR ROBUST DESIGN

Taguchi's quality engineering methods stand in stark contrast to the traditional approach described above. So different, that many engineers and engineering managers see them as radical departures from standard design practice. Figure 2 shows a flow chart of the Taguchi parameter design process including initial benchmarking, concept generation, and concept selection activities which lead into other activities such as design decomposition, ideal function development, measurement system selection, followed by parameter and tolerance design. Benchmarking of competitors' products or one's own earlier product designs, is where the difference is first apparent. Signal-to-noise ratios are evaluated for important functions, both to show how much robustness improvement needs to be made to handle predicted downstream conditions (and predicted competitor improvements,) and also to indicate when to stop further engineering development. Where benchmark data are not available, as with new technologies, closeness of the design's function to ideal function is used as a benchmark.

Engineers, given a function to be developed such as paper feeding, can generate many designs to perform this function. Taguchi recommends selecting one to three of the best concepts utilizing concept selection techniques such as those introduced by Stuart Pugh, (Pugh, 1983) and determining whether the concept's robustness to important predicted downstream conditions (i.e. expected manufacturing and usage conditions) can be quickly improved. Robustness is defined here as a design's ability to convert input energy into just desired function. Figure 3 shows the robust design cycle for comparison with the traditional design cycle shown in figure 1. Note that the loop is back to a new design concept after failure to improve the initial design's robustness, rather than loop back to a design fix when the function does not work well as nominal conditions.

If a team of engineers has used available knowledge (e.g., physics, mechanistic models, experience with similar designs) to improve the robustness of a given concept, and the robustness has not improved significantly, then the concept should be dropped, usually within one month after selection. Another concept should be selected, and the same approach applied. This process can be done concurrently with two or more small design teams evaluating alternate designs. It is most important to identify designs with robustness potential to predicted downstream noise conditions first, rather than continue to try to develop a concept that cannot be made robust even with large cost and time expenditures. Downstream noise conditions include factors that are expected to affect the design's main function, but which, for one reason or another, cannot be controlled. It is most important in this approach to develop very flexible hardware, so that large areas of design space can be explored. One needs to demonstrate an ability to quickly improve robustness using available knowledge of each design's critical factors. Orthogonal arrays and signal-to-noise ratios are used to inspect the selected designs for robustness.

Testing of the selected design concepts for robustness focuses on the main function. The main function of any engineering design is to transform energy, e.g., mechanical-to-electrical, mechanical-to-mechanical. For example, the main function of the paper feeder is to transfer mechanical work from the feed roll to the paper stack top sheet. Unfortunately, measurement of the work or energy transferred to the sheet is not possible with our present measurement technology. It is, however, possible to measure the normal force on the stack and determine the thresholds for misfeed and multifeed. These good continuous monotonic measures are related to the basic energy transfer mechanism of the design's function. The mean squared deviation from zero is the objective function to be minimized for misfeed threshold data, and the mean squared deviation from infinity is the objective function to be minimized for the multifeed threshold data.

Table A. Traditional Xerox versus Taguchi Method (Philosophy)

Traditional Xerox	Taguchi
• Scientific Emphasis--Undertanding why it works	• Engineering Emphasis--Make it work and understanding follows
• Fine tune one design concept	• Select optimum from various concepts
• Start with high cost design and cost down selectively	• Start with low cost design and cost up selectively
• Reduce or eliminate sources of variation	• Make system insensitive to sources of variation
• Stop improvement when arbitrary specification are met	• Continue improvement beyond arbitrary specifications
• Minimize Producer cost	• Minimize the sum of producer and customer loss
• Customer specification drives technology	•Technology developed for range of customer specifications
• Emphasis on problem solving	• Emphasis on problem prevention

Table B. Traditional Xerox versus Taguchi Method (Product Development Cycle Methodology)

Traditional Xerox	Taguchi
• Emphasis on system testing for inspection	• Emphasis on parameter testing for improvement
• Several system tests to fix design quality problems	• System verification test to confirm robust design
• Early manufacturing intent hardware with little flexibility	• Early hardware with maximum flexibility for optimization with manufacturing intent hardware to follow
• Measure dysfunction (e.g. jams due to arrival times)	• Measure function (e.g. arrival time distribution)
• Test one factor at a time	• Test many factors simultaneously
• Expand time and resources gathering reliability data for management information	• Expand time and resources gathering signal to noise data for engineering optimization

To minimize cost, the Taguchi method design starts with low grade materials, low cost tolerances for hardware, inexpensive components, and no closed loop control systems. Parameter design whose basic steps are shown in Figure 2, identifies the best nominal values for design factors (with no cost increase) to improve the main function robustness. Subsequent tolerance design activities identify important tradeoffs between quality and design cost. On-line quality control activities identify economically justified countermeasures for process output variability.

In Tables A and B, I have attempted to summarize the differences in both philosophy and methodology that existed between traditional product development methods and Taguchi methods.

What to Measure?

To quickly improve an engineering design, one must be able to measure something related to the intended function. Taguchi advises not measuring quality to improve quality, rather measure engineering function to improve quality. For a paper feeder, most engineers would choose to directly measure design quality, i.e., measure what the customer sees when the function fails by counting misfeeds or multifeeds and calculating failure rates. There are usually targets given by product planning for feeder shut down rate (SDR), and unscheduled maintenance (UM) rate. Engineering management, in turn, wants to know SDR, and UM rate to track performance improvements during product development. The engineer, to keep management happy, measures SDR and UMR each time the design is modified. Unfortunately, these quality measures have problems. First, they are binary data which invariably produce inferior reproducibility. Second, they are usually full of strong interactions, which means poor predictability under variable downstream conditions. Third, they are costly and require extensive testing time while the design is held fixed.

According to Taguchi Quality Engineering Method, engineers should translate quality (dysfunction) measures into a functionality measures. For example, the function of the paper feeder is not to misfeed or multifeed, rather, it is to efficiently and consistently transfer work from the feed roller to the top sheet of paper in the stack.

Most statisticians do not consider what to measure when advising engineers, and the engineer usually selects a quality characteristic to please management. As a result, the data are almost always full of strong interactions, which require nothing less than full factorial designs (all combinations) to assure good model fitting. The statistical literature has done a very poor job of helping engineers understand the nature of interactions. Interactions can be minimized by appropriate countermeasures such as selection of response, selection of signal-to-noise ratio, intelligent selection of control factors, sliding scales for control factors which are related, and others. A good engineering design is one which is interaction free because it is highly desirable to have a system for which stability gains made by setting the various control factors at their best nominal levels are additive, i.e. the superposition principle applies. Customers need to be protected from designs which are unpredictable. This interaction free design criteria is quite often overlooked in traditional design process. Consequently, a good deal of the re-engineering work results due to passing a design downstream which is full of interactions, especially those adversely affecting stability.

Interactions manifest themselves as non-additivity of effect of different design parameters. To avoid interactions, engineers need to select measures related to energy or work transfer. To the extent that you can measure something related to energy transfer, your contribution rate from interactions will be minimized. An example that Taguchi often uses to illustrate the idea is designing a resistor. Do you measure failure rate, yield, or some other dysfunction in order to optimize the design?

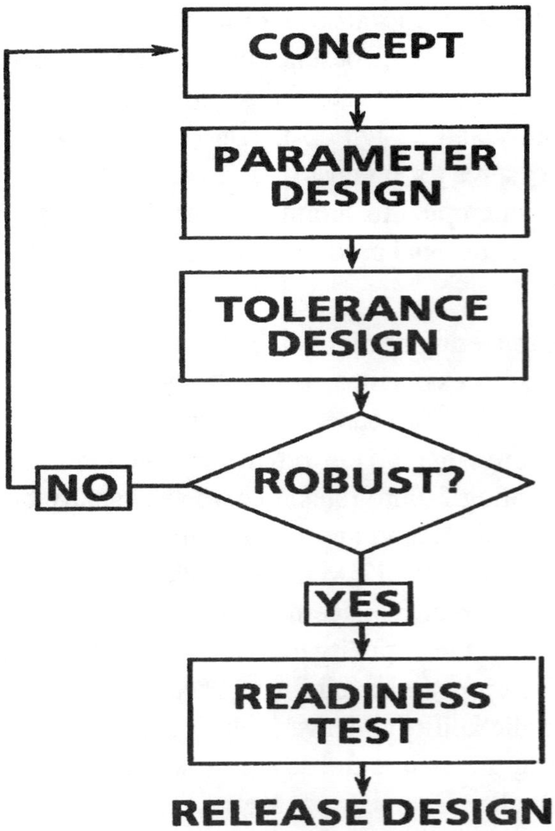

Figure 2 Robust Design Cycle

The answer is no. There is no reason to believe that any of these characteristics have additivity. Many engineers would measure resistance, but again, resistance is a quality characteristic, which may be specified after design, not an engineering function. The resistor's main function is to convert electric current into voltage. If one could design a resistor whose voltage differential was proportional to current (signal factor) over a very wide range of currents, and other downstream usage conditions, such as elevated temperatures, one would capture the entire market for resistors. Each new customer's target value for resistance could be met by simply tuning the production system with one or two adjustment factor. If you measure resistance in designing a resistor, you must have a customer's target in mind. When a different target is specified, re-optimization is usually required.

In case of the paper feeder, if the engineer were to measure lead edge arrival time at some downstream sensor, (which is related to the energy imparted to the paper), this would be a useful measure to optimize robustness so long as the paper did not misfeed or multifeed. If the paper lead edge never arrives at the sensor, however, what number does one use for arrival time? If more than one sheet arrives at the sensor, what number does one use for arrival time? Sometimes, limitations in measurement technology require a different approach than directly measuring the system's output. In the case of optimization of the paper feeder, where misfeeds and multifeeds are likely with the large design space being considered, the operating window method developed by Taguchi and Clausing is preferred. This method will be described in the next section.

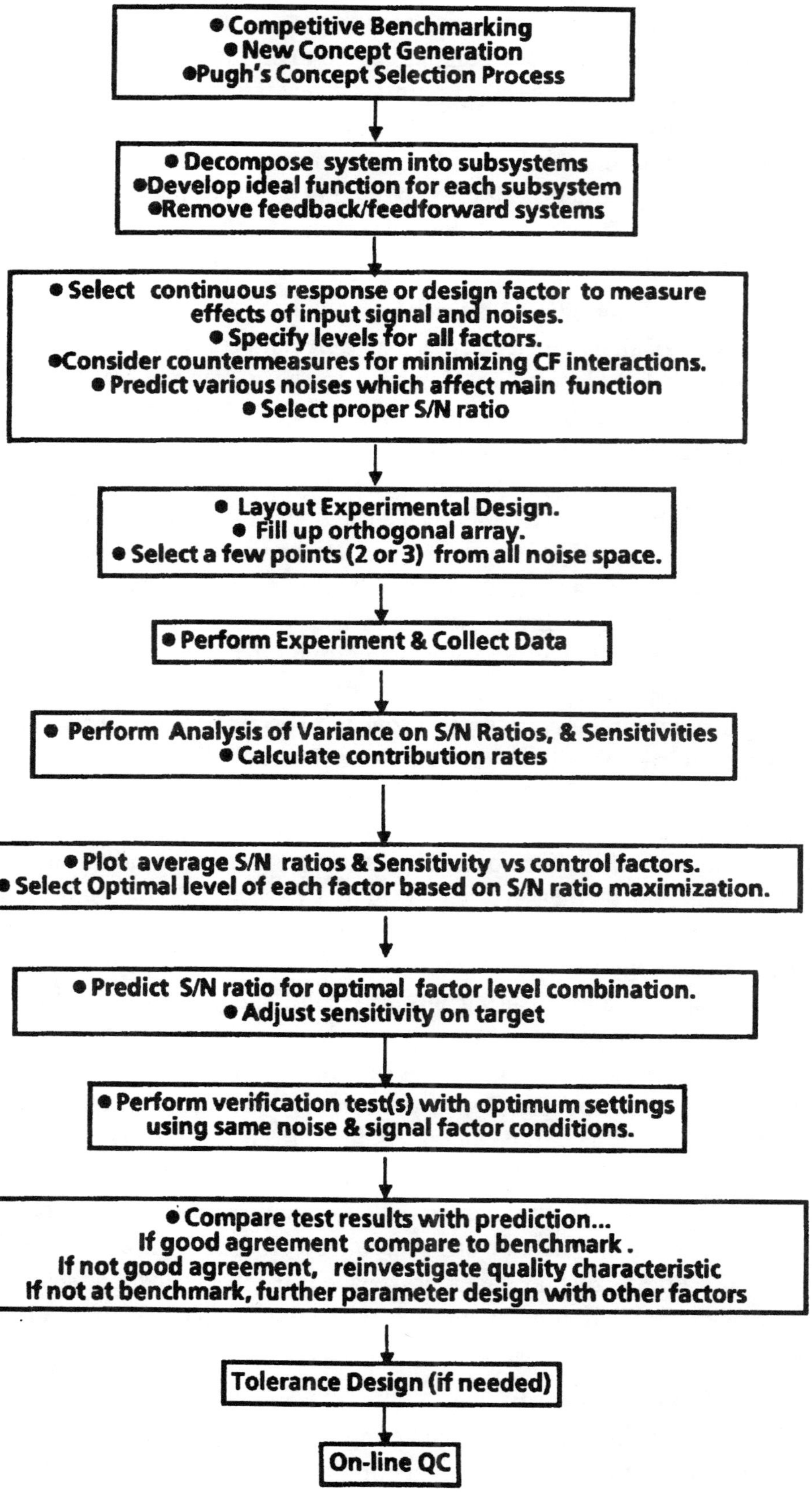

Figure 3 Parameter Design Optimization Process for Technology Development

One effective countermeasure for avoiding interactions as mentioned before, is the intelligent selection of control factors that affect the main function. For example, if the engineer knows that the ratio of two factors is the important effect rather than the individual factor levels, he should use that knowledge to avoid interactions by assigning one factor as usual and assigning the ratio of the two factors as the second. If the engineer understands the technology well enough, he should be able to make these decisions prior to running the experiment, and demonstrate the desired interaction-free stability. If you assign control factors blindly to orthogonal arrays, you will have problems with interactions.

Taguchi Experimentation Philosophy

The objective of all parameter design experimentation is to improve the main function of the design (functionability: as labeled by Taguchi) in the presence of good noise factors, not to understand the underlying mechanisms (of misfeed or multifeed). Any well designed experiment, however, should provide information about the design which can later be used to improve our understanding of the underlying phenomena. Understanding the underlying mechanisms, however, is considered a scientific objective rather than an engineering objective. In science, the objective is to create a model that expresses the system—to discover relationships that lead to models of cause and effect in order to derive laws that explain natural phenomena. Results of scientific studies are not patentable, and have only indirect effect on company profitability. In engineering, the objective is the development of products and processes at minimum cost. Cost, development time, and product stability are critical considerations. Results of engineering activities are patentable and directly affect company profitability. According to Taguchi, "Too many experiments are conducted by U.S. engineers with scientific objectives rather than engineering objectives. This inevitably extends product development time. Engineers in the U.S. are very good at inventing new concepts, and very bad at optimizing them quickly."

Taguchi considers the U. S. practice of measuring a multitude of response inefficient and wasteful. Development of measurement technology, according to Taguchi, is very important because it determines what and how quickly you are able to measure the design's main function. "Measurement technology readiness determines your overall competitive position in the market place." Taguchi recommends measuring only the system, subsystem, or component's main function. "If you can't measure what your design is supposed to do, you will not be able to improve it quickly, while the market window is still open", says Taguchi. Anyone can count jams and misfeeds. There is no competitive advantage in counting. If you can only measure what your design is not supposed to do, your development time will be increased through having to measure characteristics that are full of interactions. The only way to optimize an interactive system is to consider all combinations (full factorial) of design factors, and even that will not assure downstream predictability. By focusing on the robustness of the main function and subsequently reducing the energy needed to perform the main function, side effects such as paper jams, paper misfeeds, paper damage, image damage will be minimized. Robustness improvements, e.g., larger signal to noise ratios, are the true reliability improvements.

Technology Development

Technology development is an engineering activity to prepare for eventual product development. It has long been the desire of engineering management to make subsystem technologies "ready" for easy integration with other subsystems when product planning identifies new product opportunities. During technology development, Taguchi advocates not waiting for customer specifications to develop hardware functionability. It is most important to improve the main function over a wide range so that the technology is ready (robust) when different customers appear. All that is required for product development then is tuning the various subsystems to work in concert. Taguchi recommends not specifying reliability targets during technology or product development. These specifications force engineers to constantly measure reliability, a very time consuming activity, especially for an already robust system. Usually measuring reliability does little to improve reliability. It is merely inspection. Reliability testing requires that the design be frozen for the duration of the testing. After optimization, i.e. achieving benchmark signal-to-noise ratio, a system verification test should be run to inspect the design's performance. Life testing can be done concurrent with product launch just to make sure something has not been missed.

Good physical models should be used if they are available, according to Taguchi. If they are not available, don't spend the time and resources to build them during engineering development, when you should be focused on design robustness, cost, and delivery time. Usually, some balance is struck between analytical modelling and empirical work.

CASE: DESIGN OF THE FRICTION RETARD FEEDER

An ideal copier is one which works perfectly each time it is used by the customer, under all operating conditions, and through its intended life, with no harmful side effects. When the copier's performance deviates from its ideal performance, its quality is inferior. The ideal function of the copier, at a system level, is to make the copy look like the original, i.e., Copy=Original. It is quite easy for the customer to compare the original to the copy. Any deviation from this ideal function causes loss to the user, in the form of dissatisfaction, recopying, adjusting machine, etc. also to the supplier in terms of potential loss market share, and in varying degrees, to the rest of society. The object of engineering design is to come as close as possible to benchmark quality (or ideal function) within time and cost constraints. Figure 4 shows the symbolic quality iceberg with traditional quality loss components shown above the water line, and the more treacherous quality loss components shown below.

The traditional U.S. quality loss function is a step function. There is zero quality loss within specification limits, and once the specification limits are exceeded, there is a measurable producer cost from scrap or rework. Pivotal in Taguchi's engineering process is the idea that quality loss is better estimated with a continuous function (orthogonal expansion around an ideal value or target). Quality loss at the target is zero and continuously increases with further deviation from the target. The loss function can be estimated using available data, and can subsequently be used for making all quality decisions, e.g., tolerances on drawings, feedback control processes, upgrading equipment.

For paper feeder, quality loss comes from misfeed threshold force not being very close to zero, and from multifeed threshold force not being very close to infinity. A quadratic function is assumed in many instances where no good data are available. Figures 5a and 5b show the two single sided quadratic functions which can be estimated knowing the coordinates of at least one point on each of the curves. That points' coordinates are usually the cost of the countermeasure when the function fails in the customer's location, A_o, and the point, y_o, where 50% of the customers are expected to complain of a problem (also called LD_{50}). The curvature, or loss coefficient, k, will usually be different for misfeed and multifeed failure modes. For each copier feeder, there willbe

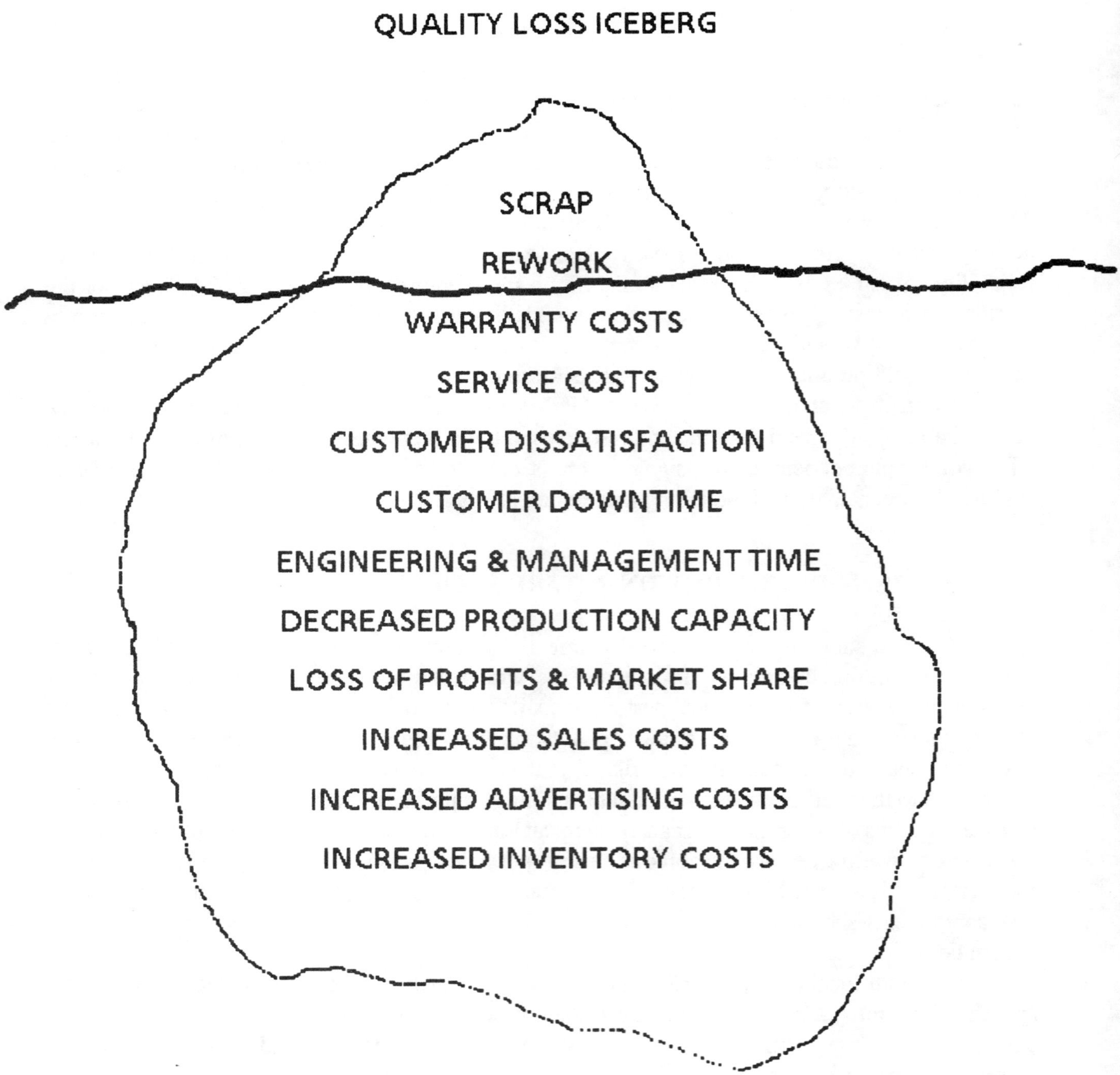

Figure 4. Quality Loss Iceberg

a quality loss associated with each one's proximity to the ideal values. The quality loss for both misfeed and multifeed is the sum of the quality loss for each failure mode, as shown in figure 5c. The quality loss for a fleet of copiers having a distribution of misfeed and multifeed thresholds can be calculated through integration. The total quality loss can also be integrated over the expected life of the program, e.g., 10 years.

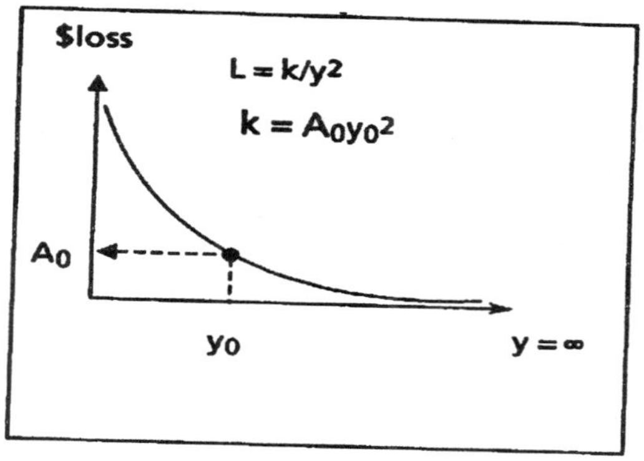

$$L = k(MSD) \; where \; MSD = \frac{1}{n}\sum_{i=1}^{n}\frac{1}{y_i^2} \approx \frac{1}{\bar{y}^2}\left(1+3\frac{\sigma^2}{\bar{y}^2}\right)$$

Figure 5a. Loss Function for Larger-the-better Characteristics

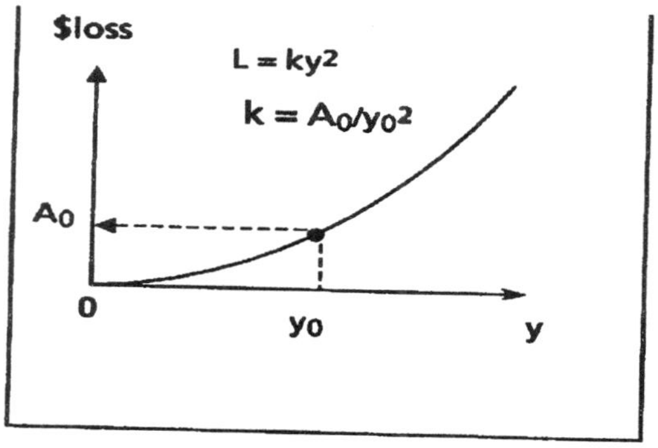

$$L = k(MSD) \; where \; MSD = \frac{1}{n}\sum_{i=1}^{n} y_i^2 = \bar{y}^2 + \sigma^2$$

Figure 5b. Loss Function for Smaller-the better Characteristics

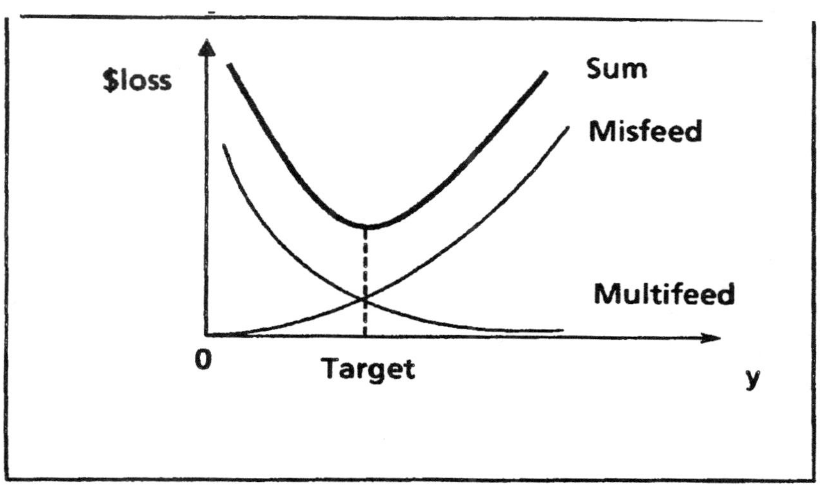

Figure 5c. Composite Loss Function for the Friction Retard Feeder

Operating Window Method

Where no dependable continuous measure of the output (functionability) of a system is available, it is advisable to select a continuous control or noise factor to improve robustness. During the planning phase of the experiment, a list of control factors and noise factors is developed which participants believe have some bearing on the functionability of the hardware. From this list, one continuous factor (control or noise) is selected. This factor will be systematically adjusted over a sufficient range to create problems. The threshold values where these problems occur, sometimes called LD_{50}, mark the operating window boundaries. More importantly, these demarcation points are determined under at least two stress conditions—the first under conditions that aggravate misfeeds (heavy weight substrates, low coefficient of friction to feed roll,...) and the second under conditions that aggravate multifeeds (light weight substrates, high static charge on substrate,...) Ideally, the misfeed threshold force should be very close to zero under all noise conditions, and the multifeed threshold force should be as large as possible under all noise conditions.

Engineers decided to use stack normal force as the control factor to adjust. Starting with zero stack force, a screw mechanism was turned in .2 lb. increments until the complete transition from no sheets fed to one sheet fed was observed. The point where 50% of the sheets fed was estimated and called the misfeed threshold. This same procedure was followed for the transition from single sheet fed to multiple sheets. In addition, both misfeed threshold values were determined under misfeed and multifeed stress conditions, respectively. The misfeed threshold values were determined under predicted customer usage conditions that moved the value away from zero stack force, and the multifeed threshold values were determined under the predicted customer usage conditions that moved the value away from infinite normal. Figure 6 shows schematically, the misfeed and multifeed transition regions for normal force under three different noise conditions which result in a small, but positive operating window. Figure 7 shows a similar picture, but with a slightly negative operating window. Negative operating window means there is no normal force setting that will enable feeding one sheet at both noise conditions.

The objective of the orthogonal array experiment is (1) to observe various alternate designs under expected customer usage conditions, and (2) to select factor levels that simultaneously move the misfeed threshold toward zero and multifeed threshold toward infinity, thereby increasing the latitude. It is important to note that it is possible to improve a design with a negative operating window. Many engineers do not appreciate this point, and choose to defer optimization until the function works under all specified conditions. This usually entails tightening tolerances and raising cost first!

The objective functions to be optimized for the paper feeder include the mean square deviation (MSD) from zero for misfeed threshold force, the MSD from infinity for the multifeed force, and finally the sum of the two. Equations 1 through 3 show the signal to noise ratios corresponding to the three objective functions. All three equations are converted to maximization problems.

Larger the better S/N ratio for multifeed threshold force: $\eta = -10\log_{10}\left[\frac{1}{n}\left(\frac{1}{y_1^2}+\frac{1}{y_2^2}+\ldots+\frac{1}{y_n^2}\right)\right]$

Smaller the better S/N for misfeed threshold force: $\eta = -10\log_{10}\frac{x_1^2 + x_2^2 + \ldots + x_n^2}{n}$

Combined S/N ratio: $\eta = -10\log_{10}\frac{x_1^2 + x_2^2 + \ldots + x_n^2}{n} - 10\log_{10}\left[\frac{1}{n}\left(\frac{1}{y_1^2} + \frac{1}{y_2^2} + \ldots + \frac{1}{y_n^2}\right)\right]$

where x_i's represent misfeed threshold values, y_i's represent multifeed threshold values. For each design configuration specified in the orthogonal array, these S/N ratios are calculated. Ultimately, the results are used to predict a new design with better stability against customer usage conditions. The optimum level for stack normal force is selected to minimize total quality loss. For symmetric misfeed and multifeed loss functions, this will be in the middle of the operating window. Naturally, all predictions of stability are verified with additional data. The verified optimum design configuration is the one which most effectively uses most of the work transferred to the paper for just feeding one sheet. Side effects, e.g., smudge, paper damage, paper dust, should be correspondingly reduced.

This methodology has been found to be quite useful for subsystems where measurement technology has not been adequately developed or where hardware has just been built. More extensive use of the methodology in early technology robustness activities is anticipated. The operating window method is not to be confused with traditional operating window method. In the traditional method, the design is mostly held fixed while one factor at a time experiments are conducted to identify the multidimensional locus of failure. After identifying the locus of failure, the coordinates of the best tuned system (to avoid failures) are identified. The traditional approach focuses on inspection and tuning the mean output of the system to avoid failures. Where there is no feasible operating point, tweaking, and tolerance tightening is usually invoked. The Taguchi approach focuses on first optimizing the design by choosing alternative nominal operating conditions to improve robustness, then tuning the system to minimize total quality loss.

Experimental Design Layout

Figures 8a and 8b show schematically one view of the paper feeder hardware, the first with the paper stack full, and the second with the stack empty. (Paper stack height was an important noise factor for the experimentation.) Figure 8b also shows several of the adjustments build into the hardware. The hardware below the feed roll was contained in a drawer which could be pulled out for easy paper tray access. Focused brainstorming meetings were held to obtain consensus on the experimental objective, what to measure, control and noise factors to include, expected outcome, contingency plans, scheduling, etc. The list of control and factors was reduced to those which could immediately be used for experimentation. The noise factors were selected based on predictions of expected customer usage conditions. The selected control and noise factors are listed in Tables 1 and 2 respectively. A total of nine control factors and five noise factors were considered. Control factors could be freely set at levels within a range determined by hardware adjustability, and available knowledge of where functional problems occurred (based on other applications and modelling work). The engineers knew that one of the critical design factors was the distance between the nip center line and the tip of the retard Mylar™ strip, and that there were two control factors, C and D that affected this distance, namely the retard pad angle and the retard Mylar™ strip length. Rather than ignoring this relationship and letting the critical design factor assume extreme levels dictated by the orthogonal array, a countermeasure called sliding scale was used.

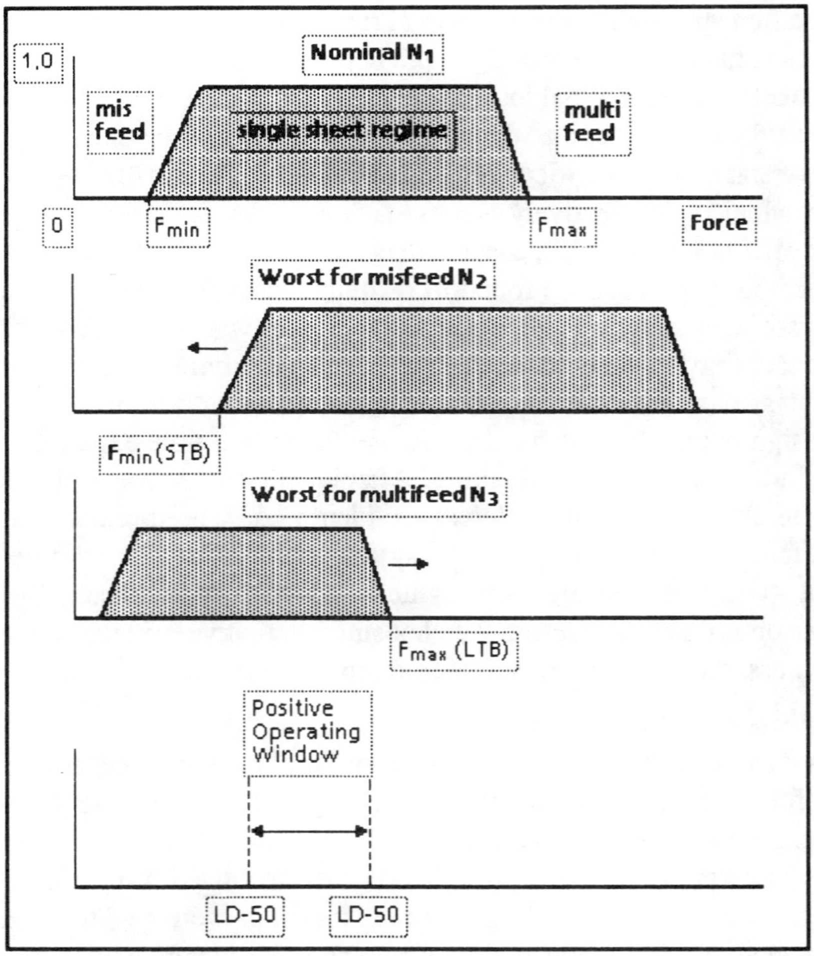

Figure 6. Effects of Three Stress Conditions, N_i on Paper Feeder Operating Window

An ideal design has zero F_{min} and infinite F_{max}

N_1 = Nominal paper

N_2 = Heavy weight paper, small friction coefficient for feed roll

N_3 = Light weight paper, high static, low RH environment

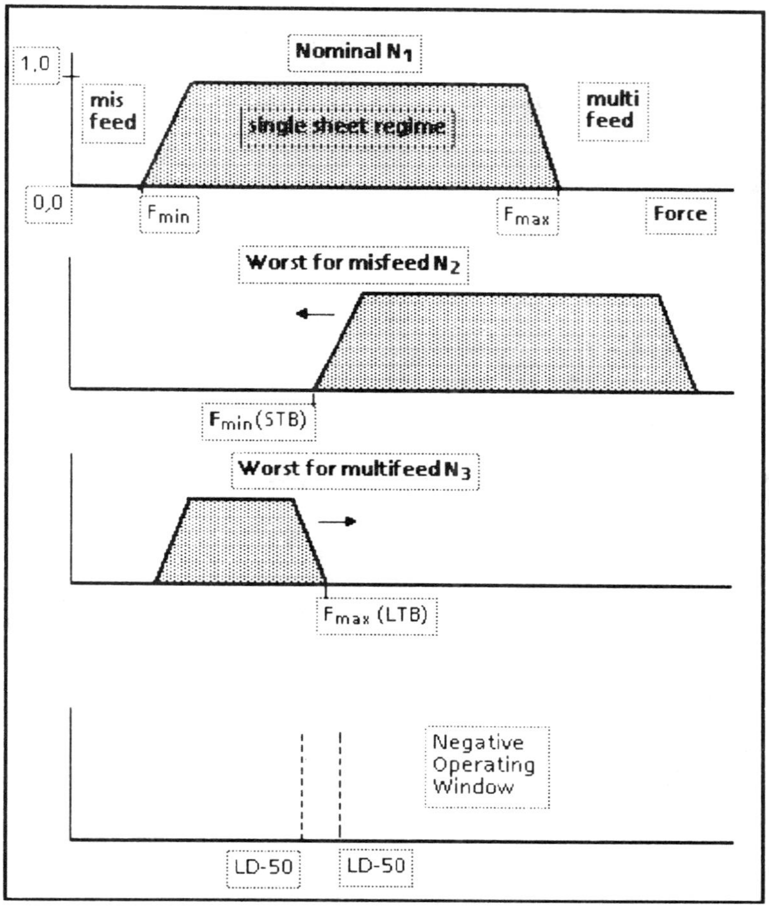

Figure 7. Effects of Three Stress Conditions, N_i on Paper Feeder Operating Window

The window can be negative, but the system can still be improved for robustness!

LD-50=median failure

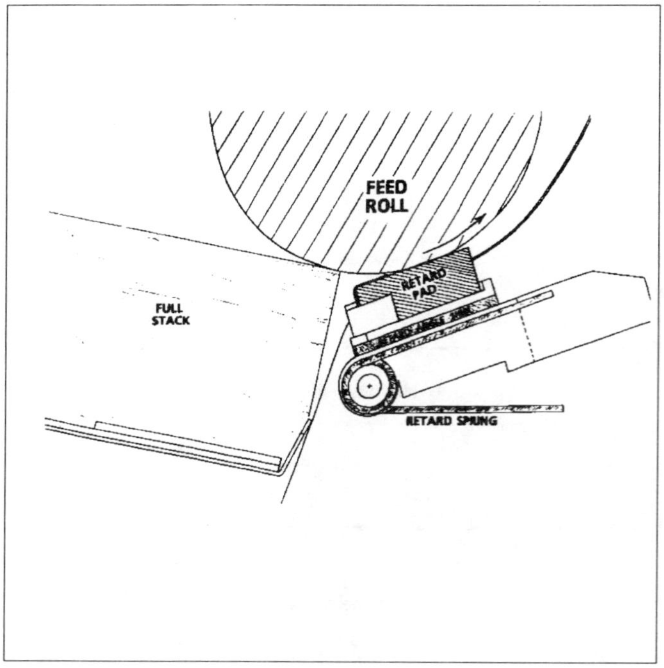

Figure 8a. Schematic View of Paper Feeder; Full Stack

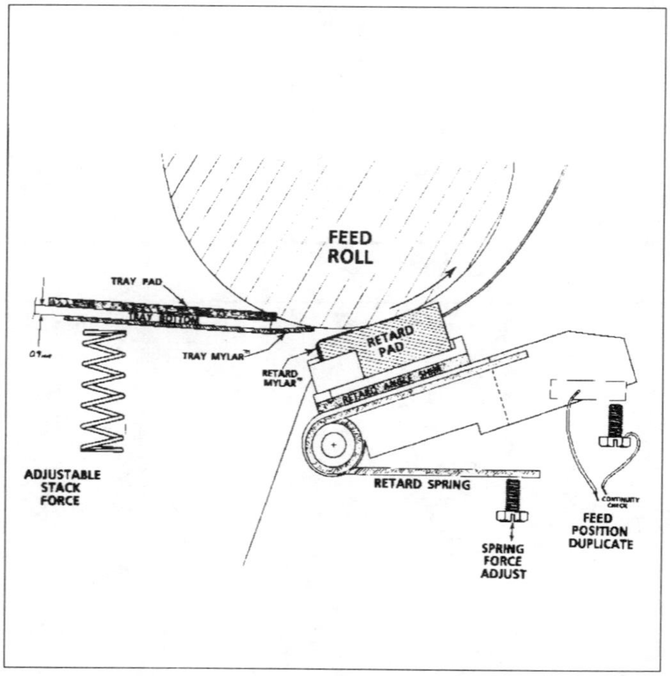

Figure 8b. Schematic View of Paper Feeder with Additional Details; Empty Stack

1 to 3 mm to give three reasonable tip-to-nip distances. With the retard pad set at 21 degrees, the range of Mylar™ strip lengths needed to be shifted up 1 mm to provide the same three levels of tip-to-nip distance. The same shift was done for the 23 degree retard pad angle.

Ignoring relationships between control factors usually introduces unwanted interactions in the data and subsequent failure to verify predicted optimum levels. It is most important to use your engineering knowledge of the design to minimize interactions. If your objective is an interaction-free design, the orthogonal array provides a means to inspect what you think you know about your particular design. If your engineering knowledge is inadequate, verification will be unlikely. If your engineering knowledge is adequate, verification will be likely and your downstream customer will be protected from operating problems. All nine control factors were assigned to an L_{18} array (2^3X3^6).

Noise factors were quite easily set, thus explaining the many sources of noise considered here. Usually, compounding of only a few stressful noise factors is done to reduce experimental effort. In this case, there was a small difference in labor involved, but a significant risk of compounding the noise factors incorrectly. Noise factors were assigned to an L_9 outer array (2^3X3^2). A more reliable feeder performanceis assured when the design is insensitive to these noise conditions.

The generic two-way layout of inner array and outer array is shown in Table 3. The 2-way layout for this paper feeder experiment is shown in Figure 9. Control factors A through H were assigned to the control factor inner array, and noise factors N through Q were assigned to the noise factor outer array. For each inner array treatment combination, nine noise factors conditions were considered. For each of the nine outer array combinations, two 10-copy runs (replicates) were made with 0.2 lb. increments of stack force. The maximum feasible range for stack force was 0.1lb. to 1.5 lb.. Above 1.5 lb force., significant distortion in the hardware occurred. This was not really a problem in that large stack force thresholds, (for multifeeds) contribute little to the larger-is-better S/N ratio i.e., reciprocal y^2. Table 4 is a form which was generated for data collection within the L_9 array. From each row of the L_9 form, the threshold values for both misfeed and multifeed (LD_{50}) were calculated.

Tables 5 and 6 are expected results from changes in both control and noise factors away from their nominal values. For example, if the retard pad angle, factor A, is reduced from the current nominal value, the experimenters expected to begin to see more frequent multifeeds. Increasing the retard pad angle above nominal should increase the frequency of mis-feeds. It is important to be able to rationalize experimental results with expected results during the orthogonal array testing. Contradictory results from any experiment should be explained.

The experimentation was completed in approximately two months using 440 work hours. This included planning, testing and data analysis. Approximately $500 was used for consumables like paper and toner.

Data Analysis Results

Table 7 contains the misfeed and multifeed threshold LD_{50} estimates for each of the OA combinations tested. For the misfeed results, where the ideal value is zero, the smaller-is-better S/N ratio is shown, and for the multifeed results, where the ideal value is infinity, the larger-is-better

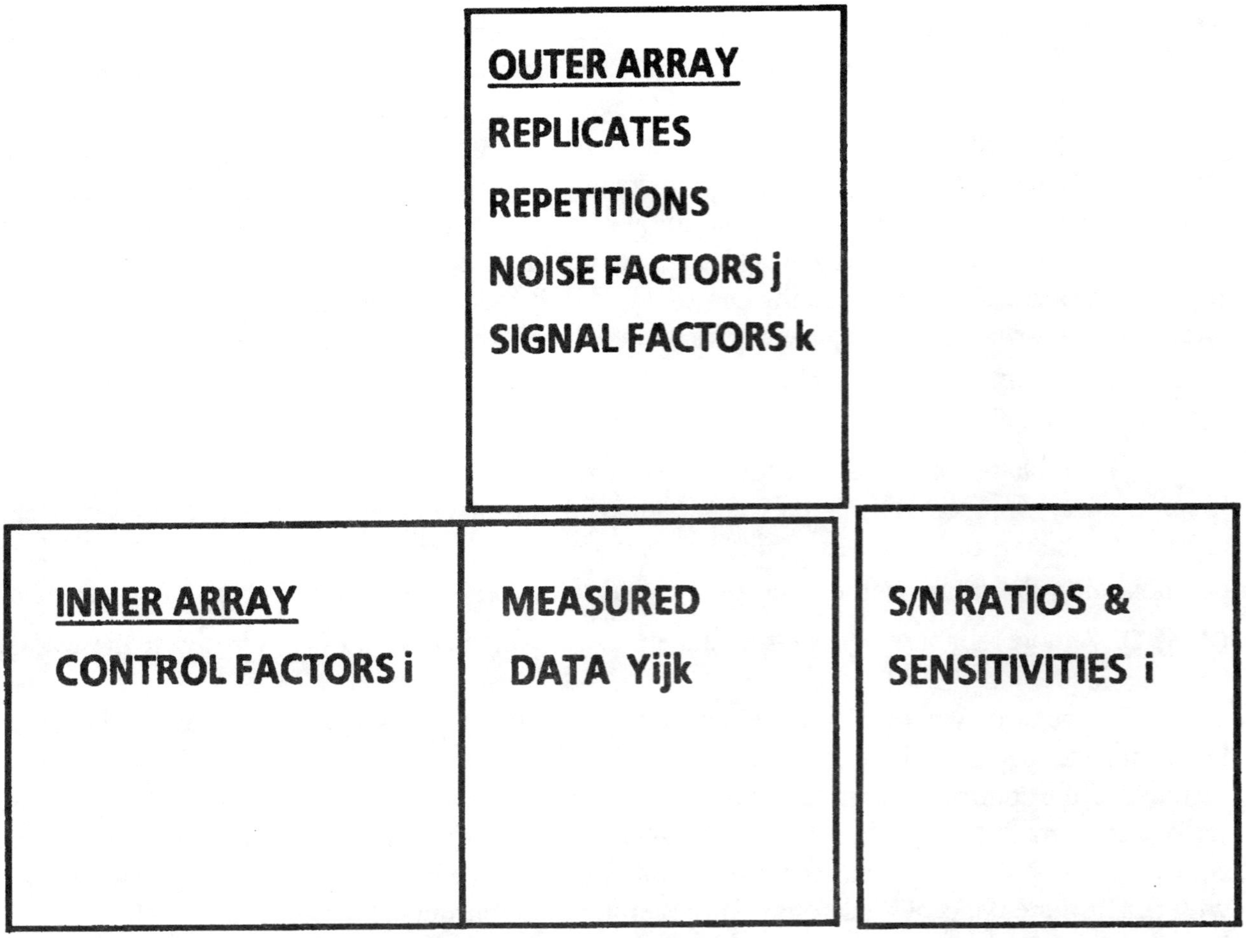

Figure 9 Generic Two-Way Layout

S/N ratio is shown. Finally, the sum of the two S/N ratios is calculated. Adding together the two S/N ratios is justified since both measures are part of the same function. Where two or more separate functions are considered, tradeoffs are usually required which means calculating average S/N ratios. A fairly large range of S/N ratios is seen for the eighteen inner array combinations, usually a sign that considerable improvement is possible. The best combination is experiment #3; the cumulative S/N ratio in the table indicates several negativeoperating window conditions, as discussed earlier.

ANOVA analysis was conducted for each S/N ratio column in Table 7 even though only the analysis of the sum of the two S/N ratios is required. Tables 8a through 8c list the ANOVA results for the three S/N ratios, respectively. Calculations of the factorial S/N ratio effects are listed in the last three columns of each table, and those results are plotted in Figures 10 through 12, respectively. In figures 10 and 11, the average stack force response is also plotted for reference. The optimum conditions for driving misfeed threshold toward zero, and driving multifeed threshold to large values is obtained by inspection of the S/N ratio plots. The best combination of misfeed and multifeed performance is obtained by similar inspection of the cumulative S/N ratio plot in Figure 12. Optimum conditions from combined S/N ratio analysis indicated the following design combination $A_1B_1C_3D_2E_1F_2G_1H_2I_3$ predicted misfeed threshold force S/N ratio for this combination was 19.6 dB with a corresponding average misfeed threshold force of 0.105 lb. Predicted multifeed threshold force S/N was 15.4 dB with an average near 5.9 lb. The sum of the two S/N ratios was 35.0 dB. The verification test was somewhat complicated by the fact that the largest stack force attainable was 1.5 lb. Nevertheless, under optimum conditions, the design was verified to have no problems with either stress noise conditions and the results showed a significant improvement over the nominal design. The nominal design, given by $A_2B_2C_2D_2E_2F_2G_2H_2I_2$, had predicted misfeed and multifeed threshold S/N ratios of 3.9dB and -3.1dB, respectively. The combined S/N ratio for the nominal design, 0.8 dB, was just slightly positive. The corresponding average misfeed and multifeed threshold forces were 0.64 and 0.70 lb., showing that the nominal design had very little latitude. The predicted nominal conditions were verified as was the corresponding S/N ratio gain in going from the nominal to the optimal design combinations. Mis-feed and multifeed failure rates in the laboratory hardware were much improved. Figure 13 graphically shows the robustness improvement predicted from these experiments.

BARRIERS TO IMPLEMENTATION OF TAGUCHI METHODS IN U. S. INDUSTRY

There have been many barriers to successful implementation of Taguchi Methods in U.S. industries. Below is a list of most probable causes:

- Misinformation about the methodology—I would estimate that 90% of the published papers I have read in the last five years contain gross inaccuracies. Many statistical experts have tried to adapt the method to their basic assumptions.
- Training programs have focused on design of experiments methods and statistical mathematics rather than design engineering principles.
- Most documented applications relate to manufacturing processes where payback for use of the methods can be much less. Consideration of robustness late in the design process, when opportunities for improvement are least and costs of design changes most, reduces appeal of using the methods.
- When engineers have little prior exposure to analysis tools which are part of the Taguchi Method (e.g., DOE, principles of design optimization methods), and/or when Taguchi Method consultants are not adequately informed about the engineering emphasis of the method, full appreciation of the method does not occur.
- Mechanistic understanding emphasis is stressed in early product development and not cost and delivery time. Many see product development as a continuation of research activities for technology development.
- Translations from Japanese publications on Taguchi methods and discussions with Taguchi suffered due to translation problems. Consequently, some of the early implementations of the methods at Xerox, for example, included glaring mistakes.
- Interfaces with the design tools (e.g., CAD/CAM, Finite Element Analysis, Quality Function Deployment, Design for Assembly and Serviceability, Value Analysis, Fault Tree Analysis, Failure Modes and Effects Analysis, Critical Parameters Development, Modeling, Statistical Process Control) are inadequate.
- Failure to document or feed back good case studies from technology development applications has been high. Institutional learning has been very slow.

ENABLERS FOR IMPLEMENTATION OF TAGUCHI METHODS

- Adoption of robustness as a design priority in technology development and early product development is required. Design reviews should focus on early design robustness and TM experts should actively participate in engineering design meetings.

• Development of a comprehensive, long range training plan including train the trainer programs with high expectations and visible support by management.

• Accurate tracking of Taguchi Methods applications, including what went right and what went wrong, and feeding back new information upstream.

• System Verification Test (SVT) both as a symbol of management's commitment to design robustness, and an assurance of benchmark competitiveness before scale up to manufacturing. The SVT will assure early focus on robustness by engineers.

• Reward and recognition parity for both problem prevention and problem solving activities.

• Periodic internal symposia to encourage TM practitioners, with sponsorship from major divisions. This allows large scale information sharing and support for further improvements.

• Benchmarking other departments or industries, including Japan, for TM implementation.

• Technology and product development processes with Taguchi Methods more fully integrated with other design activities.

• Inclusion of TM topics in engineering curricula at universities to minimize training requirements.

CONCLUDING REMARKS

Case study development has been the thrust for Taguchi Methods training and deployment for the last 8 years at Xerox (Dr. Taguchi recommends this long term approach to help engineers internalize the basic concepts). Many hundred case studies are conducted each year, but mostly downstream, ether to fix existing non-robust manufacturing processes, or to minimize service costs from continued customer complaints about non-robust subsystems. Applications downstream tend to use nonconformant Taguchi methods, for lack of good understanding. Many engineers believe that just using one of the standard orthogonal arrays, for example, constitutes conformant use of the methodology.

Middle and top levels management, by and large, have only an awareness of Taguchi's design methods. Some have exposure to the basic design philosophy, (e.g. the importance of customer satisfaction and continuous improvement) but do not fully appreciate the efforts needed to implement the methodology. There are pockets of isolated excellence in use of the methodology, but widespread proliferation can be improved by the following:

• When Taguchi Methods Training is delivered, there must be a good measure of the effectiveness or benefit of the training, otherwise the risk is high that training will not improve over time.

• There should be logical sequencing with other design tools and methods to avoid the risk of confusion.

• The training should be customized to each application to avoid the risk of lost interest.

• Continued reinforcement and planning is required to develop champions, otherwise stagnation can occur.

• Training should be targeted to groups who will actually use the information. Training given enmasse risks much wasted time.

• Training should be sufficient to minimize the risk of fatal mistakes. Such mistakes have been shown to mask the potential benefits of continued use of the methods.

• Useful information about particular design applications needs to be fed back upstream to avoid making the same mistakes.

Although the Taguchi Method has been mapped into our product delivery process, the decision to use Taguchi's methods is largely left to the discretion of the engineer. Design reviews, at critical junctures, need to address questions of robustness, so there is an incentive to focus on robustness early. In traditional design reviews a list of the top ten problems for each subsystem is presented along with plans for identifying and eliminating the root cause of the various problems. Problem solving is always more comfortable than problem prevention.

Dr. Taguchi visits Xerox biannually, during which time he works with engineers and trainers to help develop good case studies. Nearly all copier subsystems engineers have had some exposure to the methodology through either internal training session or through video tape sessions. Still, this is not enough. Other prominent statistical consultants criticize the methodology as being flawed, based on their limited understanding of design engineering problems. This leads to considerable confusion about the merits of the method and much effort is expended in reducing this confusion. It is most important that a firm assessment of the methods be developed (based on a well understood rule set describing the criteria for a good design) and documented for future planning.

The notion of incremental improvement of design is difficult for many U.S. engineers. The notion that the next design change or the next new concept will be a 'home run' is quite prevalent. There is very little incentive to use and improve old reliable technology. Speeding up the copier by just 5% may require all new hardware design for lack of earlier robustness development. Some subsystems may have as many unique designs as there are copier products. Taguchi's methods encourage the development of robust subsystems that can be integrated together into a variety of new products with no more than tuning for operation.

Acknowledgement

I would like to acknowledge many useful discussions of this work with Dr. Genichi Taguchi, and several people at Xerox including Nuriel Samuel, Bob Huffman, Bob Yax, H. Bober, M. Plain, D. Smith, D. Johnston, and B. Windship.

REFERENCES

Hauser, J. R. and Clausing, D. (1988), "The House of Quality", HBR, May-June, 1988, pp. 63-73.

Pugh, S. (1981), "Concept Selection - a Method that works," Proceedings of the International Conference on Engineering Design, Rome, March 1981.

Phadke, M. S. (1989), 'Quality Engineering using Robust Design', Prentice Hall, 1989

Sullivan, L. P. (1986), "Quality Function Deployment," Quality Progress, June 1986, p. 39.

Seshadri, V. (1988), " Application of the Taguchi Method for Facsimile Performance Charaterization on AT&T International Network." Proceedings of IEEE International Conference on Communications, Philadelphia, PA, June 1988, pp. 0833-0835.

Taguchi, G. (1977), 'Jikken Keikakuho', 3rd Edition, Tokyo, Japan: Maruzen, vol. 1 and 2, 1977 and 1978 (in Japanese). English translation: Taguchi, G. System of Experimental Design, Edited by Don Clausing. New York: UNIPUB/Kraus International Publications, vol. 1 and 2, 1987.

Taguchi, G. and Wu , Yn-In (1979), "Introduction to Off-Line Quality Control". Central Japan Quality Control Asociation. Meieki Nakamura-Ku Magaya, Japan, 1979. (Available from American Supplier Institute, Inc., Dearborn, MI).

Taguchi, G. (1987), System Experimental Design. Engineering Methods to Optimize Quality and Minimize Costs. UNIPUB/Kraus International Publications, 1987.

Figure 10 TRAY 3 FEEDER OPTIMIZATION
MISFEED DATA

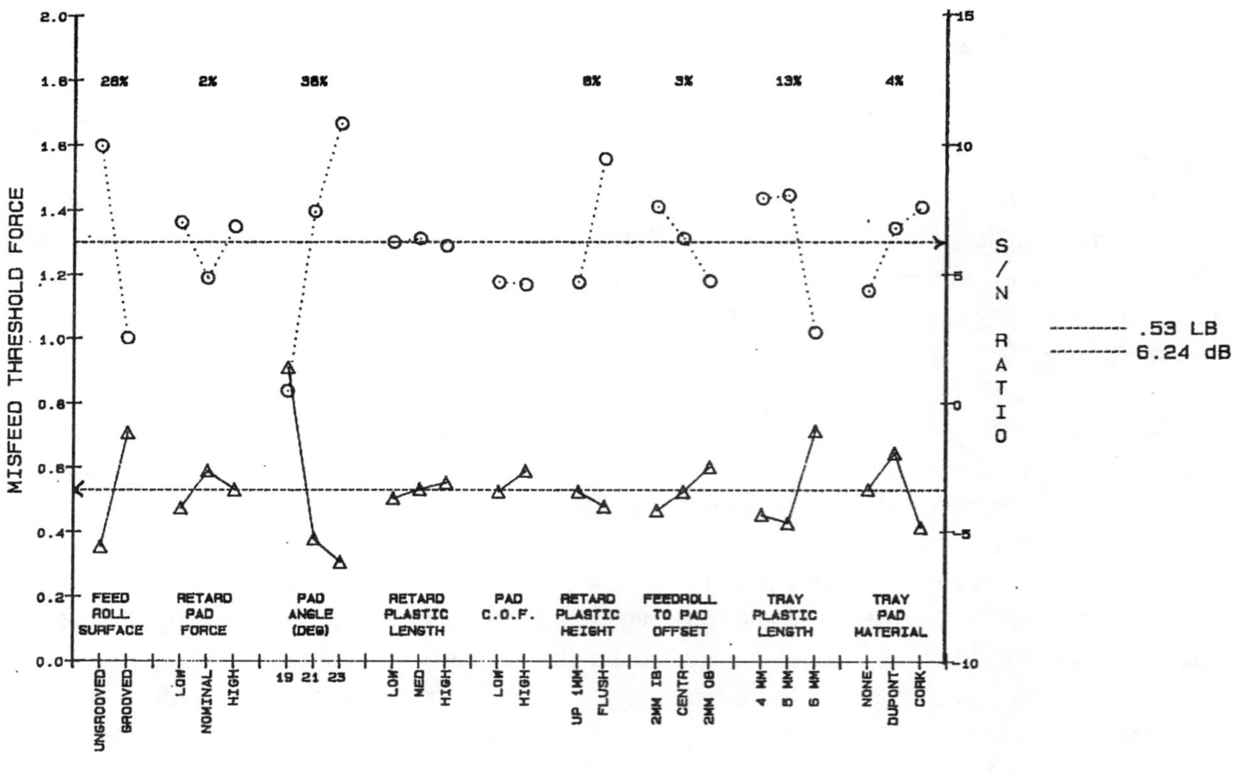

Figure 11 TRAY 3 FEEDER OPTIMIZATION
SUM OF TWO S/N RATIOS
MISFEED+MULTIFEED

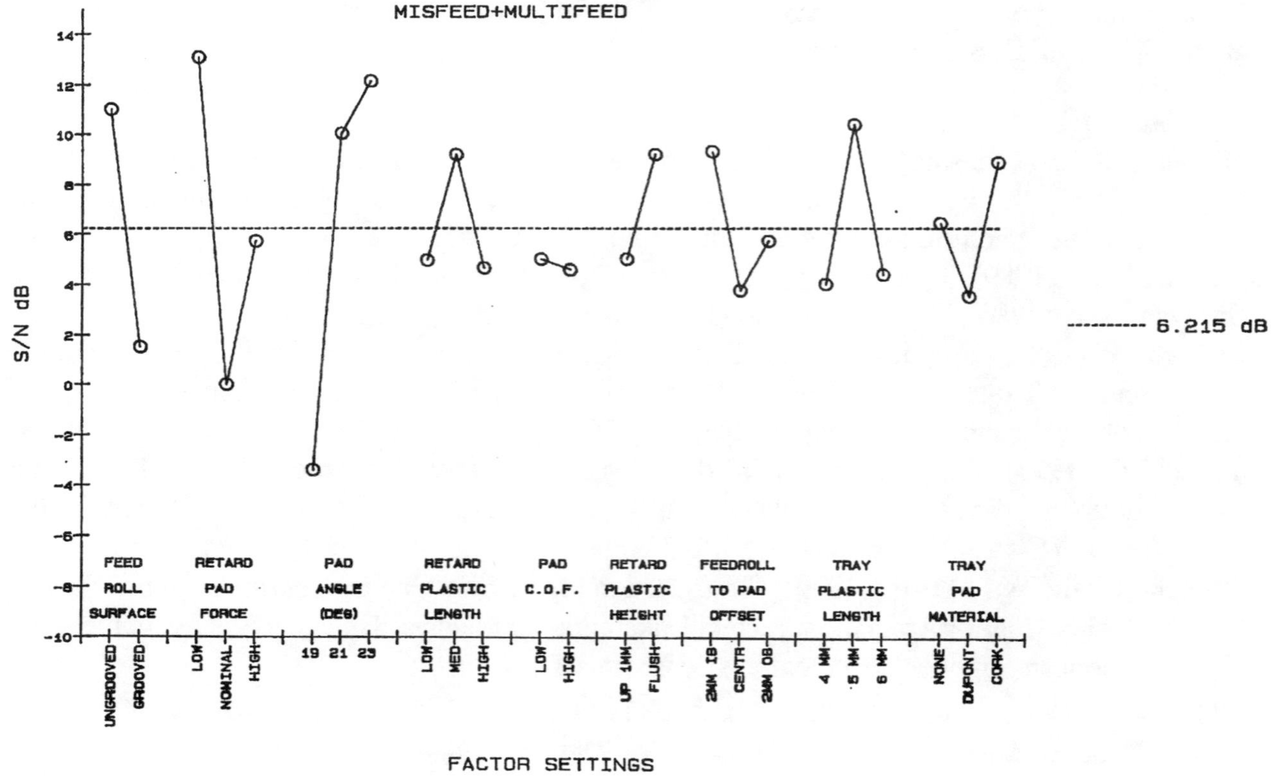

Figure 12 TRAY 3 FEEDER OPTIMIZATION MULTIFEED DATA

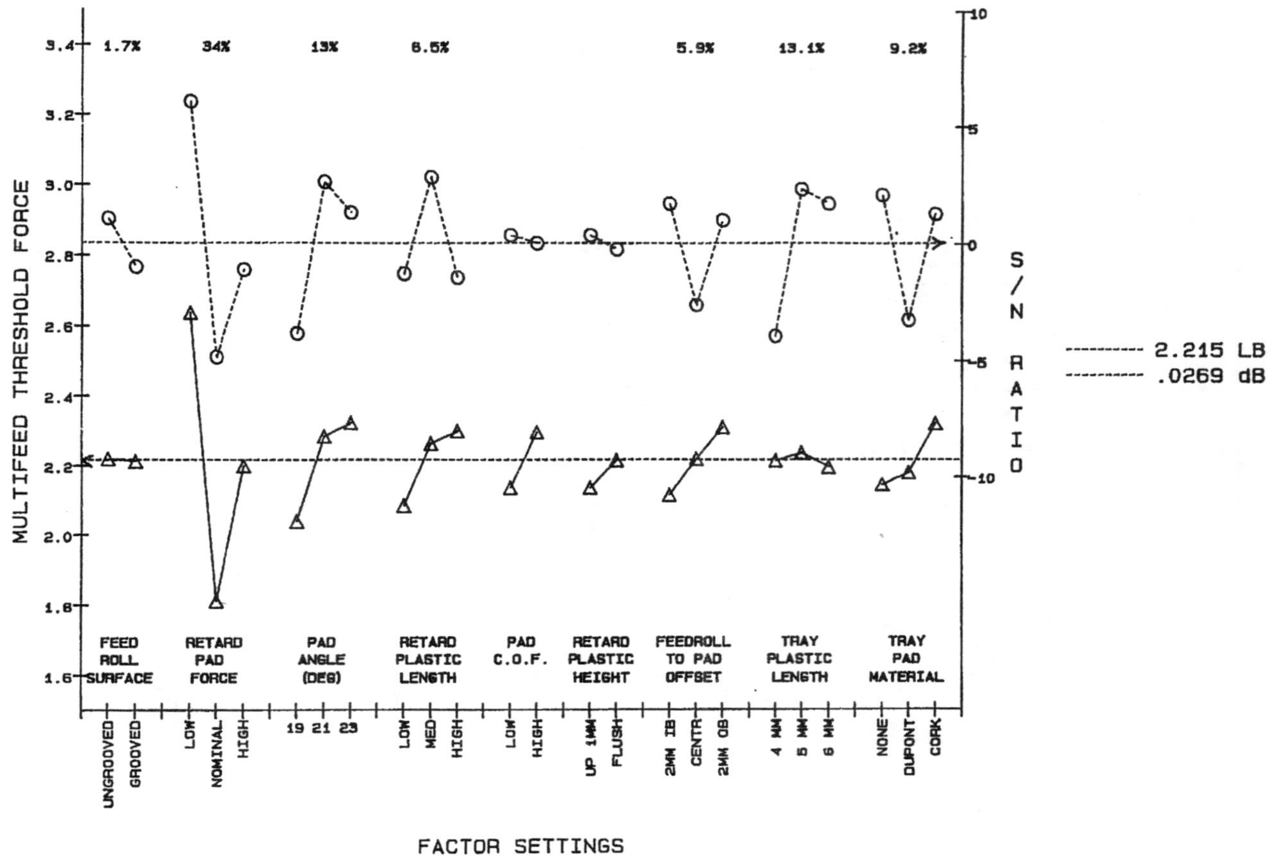

Opening the Operating Window through Robust Design

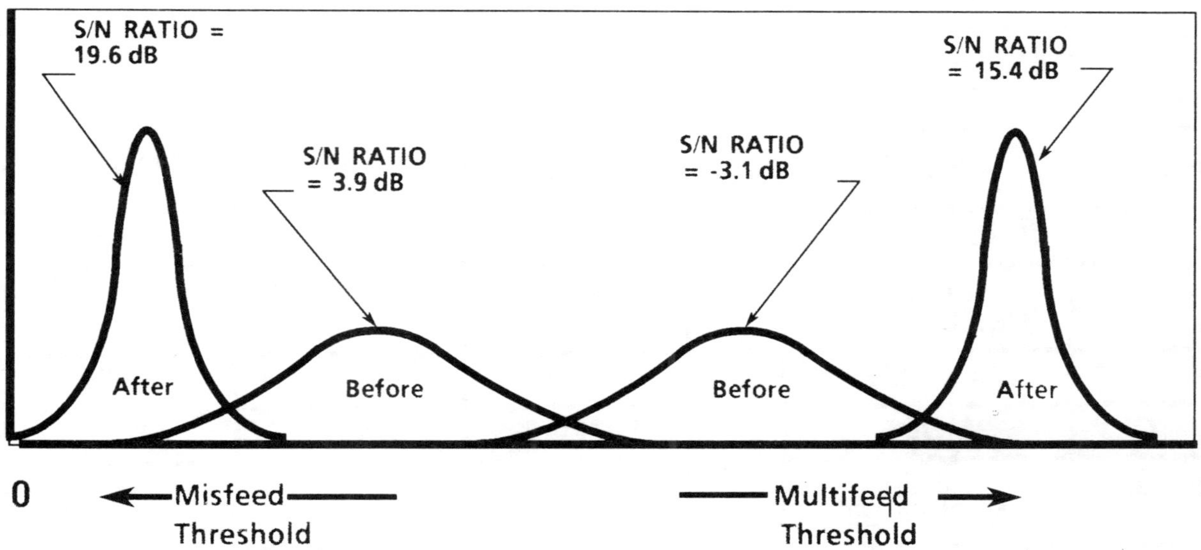

Figure 13 **PAPER FEEDER DISTRIBUTIONS**

Table 1 Control Factors

Label		COL	LEVEL 1	LEVEL 2	LEVEL 3
A	FEED ROLL CONFIGURATION	1	UNGROOVED	**GROOVED**	
B	RETARD PAD FORCE	2	LOW	**NOMINAL 5.6 MM**	HIGH
C	RETARD PAD ANGLE (DEG)	3	19	**21**	23
D	RETARD MYLAR™ LENGTH (MM)	4	1 2 3	**2** **3** **4**	3 4 5
E_L	PAD COEFFICIENT OF FRICTION	5	LOW	**HIGH**	
E_R	MYLAR™ HEIGHT ABOVE PAD	5	UP 1MM	**FLUSH**	
F	FEED ROLL TO PAD LATERAL OFFSET	6	-2MM	**CENTERED**	+ 2MM
G	TRAY MYLAR™ LENGTH (MM)	7	4	**5**	6
H	TRAY PAD MATERIAL	8	NONE	**PORON™**	CORK

Table 2 Noise Factors

Label	NAME	COL	LEVEL 1	LEVEL 2	LEVEL 3
N	PAPER WEIGHT	1	16#	20#	32#
O	PAPER ORIENTATION	2	SHORT EDGE FEED	LONG EDGE FEED	
P	STACK HEIGHT	3	LOW	MED	HIGH
Q_L	MODE	4	SIMPLEX	DUPLEX	
Q_R	REAM INTERFACES	4	YES	NO	

Table 3
Friction Retard Feeder Optimization
L_{18} x L_9 (Two-way Layout)

No.	Control Factor Array	Noise Factor Array L_9 N 1 1 1 2 2 2 3 3 3 O 1 2 3 1 2 3 1 2 3 P 1 2 3 2 3 1 3 1 2 Q 1 2 3 3 1 2 2 3 1	Misfeed S/N Ratio* (dB)	Multi feed S/N Ratio† (dB)	S/N Ratio Sum
	A B C D E F G H	Misfeed Threshold Data Multifeed Threshold Data			
1	1 1 1 1 1 1 1 1				
2	1 1 2 2 2 2 2 2				
3	1 1 3 3 3 3 3 3				
4	1 2 1 1 2 2 3 3				
5	1 2 2 2 3 3 1 1				
6	1 2 3 3 1 1 2 2				
7	1 3 1 2 1 3 2 3				
8	1 3 2 3 2 1 3 1				
9	1 3 3 1 3 2 1 2				
10	2 1 1 3 3 2 2 1				
11	2 1 2 1 1 3 3 2				
12	2 1 3 2 2 1 1 3				
13	2 2 1 2 3 1 3 2				
14	2 2 2 3 1 2 1 3				
15	2 2 3 1 2 3 2 1				
16	2 3 1 3 2 3 1 2				
17	2 3 2 1 3 1 2 3				
18	2 3 3 2 1 2 3 1				

* S/N Ratio (SIB) = $-10 \log_{10} [1/n \times \Sigma y_i^2]$

† S/N Ratio (LIB) = $-10 \log_{10} [1/n \times \Sigma y_i^{-2}]$

Table 4 Data Collection Form for L_9 Outer Array

EXPER. # ___1___

DATE ________

MACH # ________

CONTROL FACTORS

FEED ROLL CONFIG.	RETARD PAD FORCE	RETARD PAD ANGLE	RETARD MYLAR™ LENGTH	RETARD PAD C.O.F. & HEIGHT	FD. ROLL TO PAD OFFSET	TRAY MYLAR™ LENGTH	TRAY PAD CONFIG
____	____	____	____	____	____	____	____

NOISE FACTOR	DESCRIPTION	Fs =	.10#	Fs =	.30#	Fs =	.50#	Fs =	.70#	Fs =	.90#	Fs =	1.1#	Fs =	1.3#	Fs =	1.5#
#	10 SHEETS / RUN 2 RUNS / CELL	RUN # 1	RUN # 2	RUN # 1	RUN # 2	RUN # 1	RUN # 2	RUN # 1	RUN # 2	RUN # 1	RUN # 2	RUN # 1	RUN # 2	RUN # 1	RUN # 2	RUN # 1	RUN # 2
1	16# - SEF - LOW STK SIMPLEX - NO INTERFACE																
2	16# - LEF - MID STK DUPLEX																
3	16# - SEF - HIGH STK SIMPLEX - WITH INTERFACE																
4	20 # - SEF - MID STK SIMPLEX - WITH INTERFACE																
5	20 # - LEF - HIGH STK SIMPLEX - NO INTERFACE																
6	20 # - SEF - LOW STK DUPLEX																
7	32 # - SEF - HGH STK DUPLEX																
8	32 # - LEF - LOW STK SIMPLEX - WITH INTERFACE																
9	32 # - SEF - MID STK SIMPLEX - NO INTERFACE																

Table 5

PREDICTED EFFECTS OF CONTROL FACTOR CHANGES

LBL	CONTROL FACTORS	< Nominal	> Nominal
C	Retard Pad Angle	Multifeed	Misfeed
B	Retard Pad Force	better	Worse
E_R	Retard Mylar™ Height	Misfeed	Multifeed
E_L	Retard Pad Coef of Friction*	Multifeeds	Better
G	Tray Mylar™ length	Misfeed	Multifeed
H	Tray Pad Coef of friction	Multifeed	Misfeed
A	Feed Roll Coef of friction	Misfeed	Better
F	IB-OB Feed Roll to Pad alignment	Worse	Better
D	Retard Mylar™ Length	Misfeed	Multifeed

Table 6

PREDICTED EFFECTS OF NOISE FACTOR CHANGES

LBL	NOISE FACTORS	< Nominal	> Nominal
N	Paper Weight	Misfeeds	Multifeeds
O_L	Paper Orientation	?	?
O_R	Stack Integrity	good	Multifeeds
P	Stack Height	misfeeds	good
Q_L	Copy Mode (Simp,Dup)	?	?
Q_R	Ream Interfaces	good	multifeeds

Table 7

Friction Retard Feeder Optimization

L_{18} x L_9 (2 way Layout)

No.	Control Factor Array	Noise Factor Array L_9 N 1 1 1 2 2 2 3 3 3 O 1 2 3 1 2 3 1 2 3 P 1 2 3 2 3 1 3 1 2 Q 1 2 3 3 1 2 2 3 1	Misfeed S/N Ratio* (dB)	Multi feed S/N Ratio† (dB)	S/N Ratio Sum
	A B C D E F G H	Misfeed Threshold Data Multifeed Threshold Data			
1	1 1 1 1 1 1 1 1	.4 .2 .6 .75 .8 .4 .8 .3 .1	5.29		6.31
		.8 1. 3. .55 3. 1.2 3. 3. 3.		1.02	
2	1 1 2 2 2 2 2 2		13.63	7.84	21.47
3	1 1 3 3 3 3 3 3		15.39	9.54	24.93
4	1 2 1 1 2 2 3 3		-.38	-4.55	-4.93
5	1 2 2 2 3 3 1 1		11.98	4.51	16.50
6	1 2 3 3 1 1 2 2		15.74	1.14	17.21
7	1 3 1 2 1 3 2 3		3.78	.37	4.15
8	1 3 2 3 2 1 3 1		4.73	3.09	7.82
9	1 3 3 1 3 2 1 2		19.53	-13.54	6.00
10	2 1 1 3 3 2 2 1		-.04	2.06	2.02
11	2 1 2 1 1 3 3 2		-2.21	6.86	4.65
12	2 1 3 2 2 1 1 3		10.10	8.98	19.08
13	2 2 1 2 3 1 3 2		-3.46	-11.40	-14.85
14	2 2 2 3 1 2 1 3		3.29	-14.02	-10.73
15	2 2 3 1 2 3 2 1		1.92	-4.80	-2.88
16	2 3 1 3 2 3 1 2		-2.49	-10.50	-12.98
17	2 3 2 1 3 1 2 3		13.31	7.30	20.61
18	2 3 3 2 1 2 3 1		2.22	6.57	8.79

* S/N Ratio (SIB) $= -10 \log_{10} [1/n \times \Sigma y_i^2]$

† S/N Ratio (LIB) $= -10 \log_{10} [1/n \times \Sigma y_i^{-2}]$

TABLE 8a

ANOVA of S/N Ratio for Misfeed Threshold Force

Source	Sum of Squares	df	% Contribution	level 1 average dB	Level 2 Average dB	Level 3 Average dB
A	249.83	1	28.4	9.97	2.52	
B	17.51	2	2.0	7.03	4.85	6.85
C	335.65	2	38.1	.45	7.45	10.82
D	.22	2	0.0	6.24	6.38	6.10
E	92.94		10.6	4.68	4.58	9.45
E_L	.03	1	0.0	4.68	4.58	
E_R	68.27	1	7.8	4.68	9.45	
F	25.20	2	2.9	7.62	6.37	4.73
G	111.96	2	12.7	7.95	8.06	2.71
H	34.04	2	3.9	4.35	6.79	7.58
Residual	13.58	2	1.5			
TOTAL	880.93	17	100			
Grand Avg.						6.24

TABLE 8b
ANOVA of S/N Ratio for Multifeed Threshold Force

Source	Sum of Squares	df	% Contribution	level 1 average dB	Level 2 Average dB	Level 3 Average dB
A	17.9	1	1.7	1.00	-1.00	
B	369.3	2	34.4	6.05	-4.85	-1.18
C	136.8	2	12.7	-3.83	2.53	1.32
D	71.2	2	6.6	-1.28	2.81	-1.52
E	1.05		.1	.33	-.06	-.25
E_L		1		.33	-.06	
E_R		1		.33	-.25	
F	62.5	2	5.8	1.62	-2.61	1.00
G	140.3	2	13.1	-3.92	2.32	1.62
H	97.7	2	9.1	2.00	-3.26	1.27
Residual	177.7	2	16.5			
TOTAL	1074.5	17	100			
Grand Avg.						0.00

TABLE 8c
ANOVA of S/N Ratio Sum for Misfeed and Multifeed Threshold Force

Source	Sum of Squares	df	% Contribution	level 1 average dB	Level 2 Average dB	Level 3 Average dB
A	408.5	1	16.2	11.05	1.52	
B	511.6	2	20.3	13.08	.05	5.73
C	854.8	2	33.9	-3.38	10.05	12.19
D	76.02	2	3.0	4.96	9.19	4.72
E	77.1		3.1	5.06	4.59	9.20
E_L	.6	1		5.06	4.59	
E_R	51.4	1	2.0	5.06	9.20	
F	96.7	2	3.8	9.36	3.77	5.73
G	154.9	2	6.1	4.03	10.43	4.40
H	83.4	2	3.3	6.42	3.58	8.85
Residual	259.1	2	10.3			
TOTAL	2522.1	17	100			
Grand Avg.						6.29

TUNING COMPUTER SYSTEMS FOR HIGH PERFORMANCE

Madhav S. Phadke*

SUMMARY

The performance of complex systems (such as computer), which have tunable parameters, can often be improved through better system administration. Tuning is necessary because the parameter settings specified by the system manufacturer pertain to specific assumptions regarding the applications and load. However, the actual applications and load could be different from those envisioned by the manufacturer.

The Robust Design Method was applied to optimize the performance of VAX/UNIX computer system. Load variation during use of the computer, from day to day and as a function of the time of day, constitutes the main noise factor for the computer system. The number of users logged on, CPU demand, I/O demand, and memory demand are some of the more important load measures.

Running Robust Design experiments on live systems involves a methodological problem that the noise conditions (which are the load conditions in the computer response time case study) are not the same for every row of the matrix experiment. Consequently, the estimated error variance can be large compared to the estimated factor effects. Also, the predicted response under the optimum conditions can have large variance. Indeed, this was observed in the present case study. Standardized S/N ratios can be used to the adverse effect of changing noise conditions, which is encountered in running Robust Design experiments on live systems. Data analysis indicated that levels of four control factor (A, D, E and F) should be changed and that the levels of the other four factors should be kept the same. It also indicated that the next round of experiments should consider setting sticky bits on more than eight commands. The verification experiment showed a 61-percent improvement in both the mean and rms response time for the standard command. This corresponds to an 83-percent reduction in variance of the response time. It also showed a 71-percent improvement in mean response time and a 79-percent improvement in rms response time for the trivial command.

---

* This case study was performed by T. W. Pao, C.S. Sherrerd, and Madhav S. Phadke at AT&T Bell Labs. Currently the author is with Phadke Associates, Inc. This paper is adapted from the author's book, *Quality Engineering Using Robust Design,* Prentice Hall, 1989.

INTRODUCTION

A computer manufacturer invariably specifies the "best" operating conditions for a computer. Why then is tuning a computer necessary? The answer lies in the fact that manufacturers specify conditions based on assumptions about applications and loads. However, actual applications and load conditions might be different or might change over a period of time, which can lead to inferior computer performance. Assuming a system has tunable parameters, two options are available to improve performance: (1) buying more hardware or (2) finding optimum settings of tunable parameters. One should always consider the option of improving the performance of a computer system through better administration before investing in additional hardware. In fact, this option should be considered for any system that has tunable parameters by the prevailing trial-and-error method may, however, prove to be excessively time consuming and expensive. Robust Design provides a systematic and cost efficient method of experimenting with a large number of tunable parameters, thus yielding greater improvements in performance quicker.

This paper presents a case study to illustrate the use of the Robust Design method in tuning computer performance. A few details have been modified for pedagogic purposes. The case study was performed by Pao, T. W., Sherrerd, C. S , and Phadke, M. S.(1985) who are considered to be the first to conduct such a study to optimize a hardware-software system using the Robust Design method.

PROBLEM FORMULATION

The case study concerns the performance of a VAX* 11-780 machine running the UNIX operating system, Release 5.0. The machine had 48 user terminal ports, two remote job entry links, four megabytes of memory, and five disk drives. The average number of users logged on at a time was between 20 and 30.

Before the start of the project, the users' perceptions were that system performance was very poor, especially in the afternoon. For an objective measurement of the response time, the experimenters used two specific representative commands called *standard* and *trivial*. The *standard* command consisted of creating, editing, and removing a file; the *trivial* command was the UNIX system *date* command, which does not involve input/output (I/O). Response times were measured by submitting these commands via the UNIX system *crontab* facility and clocking the time taken for the computer to respond using the UNIX system *timex* command, both of which are automatic system processes.

For the particular users of this machine, the response time for the *standard* and *trivial* commands could be considered representative of the response time for other various commands for that computer. In some other computer installations, response time for the compilation of a C program or the time taken for the *troff* command (a text processing command) may be more representative.

Figure 1(a)-(b) shows the variation of the response times as functions of time of day for the standard and trivial commands. Note that at the start of the study, the average response time increased as the afternoon progressed (see the curves marked "Initial" in the figure). The increase in response time correlated well with the increase in the work load during the afternoon. The objective in the experiment was to make the response time uniformly small throughout the day, even when the load increased as usual.

There are two broad approaches for optimizing a complex system such as a computer: (1) micro-modeling and (2) macro-modeling. They are explained next.

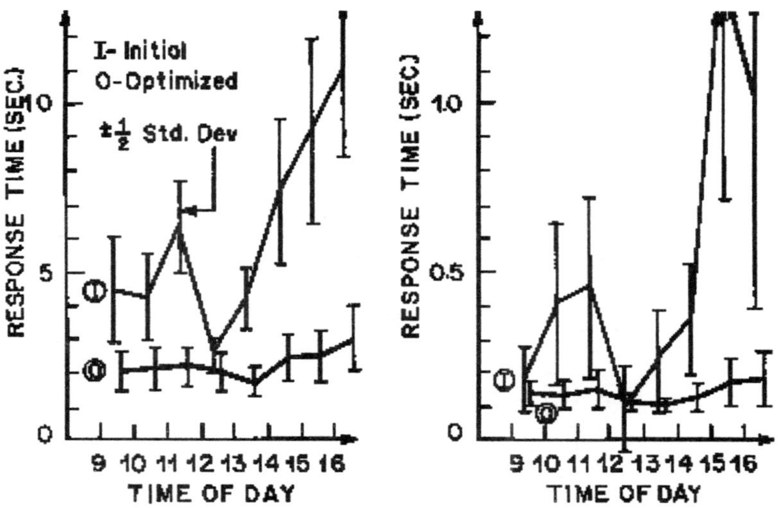

(a) for *Standard* Command (b) For the *Trivial* Command

Figure 1 Comparison of Response Times

Micro-Modeling

As the name suggests, micro-modeling is based on an in-depth understanding of the system. It begins by developing a mathematical model of the system, which, in this case, would be of the complex internal queuing of the UNIX operating system. When systems are complex, as in the case study, we must make assumptions that simplify the operation, as well as put forth considerable effort to develop the model. Furthermore, the more simplifying we do, the less realistic the model will be, and hence, the less adequate it will be for precise optimization. But once an adequate model is constructed, a number of well-known optimization methods, including Robust Design, can be used to find the best system configuration.

Macro-Modeling

In macro-modeling, we bypass the step of building a mathematical model of the system. Our concern is primarily with obtaining the optimum system configuration, not with obtaining a detailed understanding of the system itself. As such, macro-modeling gives faster and more efficient results.

It gives the specific information needed for optimization with a minimum expenditure of experimental resources.

The UNIX system is viewed as a "black box," as illustrated in Figure 2. The parameters that influence the response time are identified and divided into two classes: noise factors and control factors. The best settings of the control factors are determined through experiments. Thus, the Robust Design method lends itself well for optimization through the macro-modeling approach.

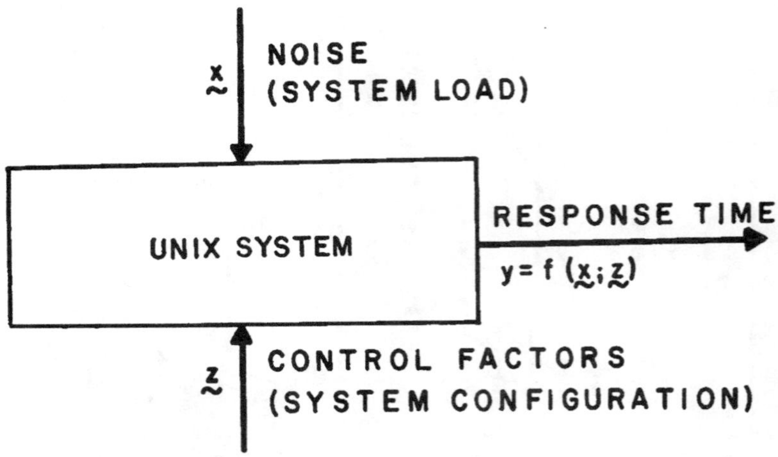

Figure 2 Block Diagram for the UNIX System

NOISE FACTORS AND TESTING CONDITIONS

Load variation during use of the machine, from day-to-day and as a function of the time of day, constitutes the main noise factor for the computer system under study.

The number of users logged on, central processor unit (CPU) demand, I/O demand, and memory demand are some of the more important load measures. Temperature and humidity variations in the computer room, as well as fluctuations in the power supply voltage, are also noise factors but are normally of minor consequence.

The case study was conducted live on a computer. As a result, the normal variation of load during the day provided the various testing conditions for evaluating the S/N/ ratio.

At the beginning of the study, the researchers examined the operating logs for the previous few weeks to evaluate the day-to-day fluctuation in response time and load. The examination revealed that the response time and the load were roughly similar for all five weekdays. This meant that Mondays could be treated the same as Tuesdays, etc. If the five days of the week had turned out to be markedly different from each other, then those differences would have had to be taken into account in planning the experiment.

QUALITY CHARACTERISTICS AND S/N RATIO

Let us first consider the standard command used in the study. Suppose it takes t_0 seconds to execute that command under the best circumstances-that is, when the load is zero, t_0 is the minimum possible time for the command. Then, it becomes obvious that the actual response time for the standard command minus t_0 is a quality characteristic that is always nonnegative and has a target value of zero-that is, the actual response time minus t_0 belongs to the smaller--the-better type problems. In the case study, the various measurements of response time showed that t_0 was much smaller than the observed response time. Hence, t_0 was ignored and the measured response was treated as a smaller-the-better type characteristic. The corresponding S/N/ ratio to be maximized is

$$\eta = -10\log_{10}(\text{mean square response time for the } \textit{standard} \text{ commands})............(1)$$

Referring to Figure 1, it is clear that at the beginning the standard deviation of the response time was large, so much so that it is shown by bars of length $\pm 1/2$ standard deviation, as opposed to the standard practice of showing ± 2 standard deviations. From the quadratic loss function considerations, reducing both the mean and variance is important. It is clear that the S/N ratio in Equation (1) accomplishes this goal because mean square response time is equal to sum of the square of the mean and the variance.

For the response time for the trivial command, the same formulation was used. That is, the S/N ratio was defined as follows:

$$\eta = -10\log_{10}(\text{mean square response time for the } \textit{trivial} \text{ commands})...............(2)$$

CONTROL FACTORS AND THEIR ALTERNATIVE LEVELS

The UNIX operating system provides a number of tunable parameters, some of which relate the hardware and others to the software. Through discussions with a group of system administrators and computer scientists, the experiment team decided to include the eight control factors listed in Table 1 for the tuning study. Among them, factors A,C, and F are hardware related, and the others are software related. A description of these parameters and their alternative levels is given next. The discussion about the selection of levels is particularly noteworthy because it reveals some of the practical difficulties faced in planning and carrying out Robust Design experiments.

The number and type of *disk drives* (factor A) is an important parameter that determines the I/O access time. At the start, there were four RM05 disks and one RP06 disk. The experimenters wanted to see the effect of adding one more RP06 disk (level A_2), as well as the effect of adding one RP07 disk and a faster memory controller (level A_3). However, the RP07 disk did not arrive in time for the experiments. So, level A_3 was defined to be the same as level A_2 for Factor A. The next section discusses the care taken in planning the matrix experiment, which allowed the experimenters to change the plan in the middle of the experiments.

Table 1 Control Factors and Levels

Factor	1	2	3
A. Disk drives (RM05 & RP06)	<u>4&1</u>	4&2	4&2
B. File distribution	<u>a</u>	b	c
C. Memory size (MB)	<u>4</u>	3	3.5
D. System buffers	1/5	1/4	<u>1/3</u>
E. Sticky bits	<u>0</u>	3	8
F. KMC's used	2	0	
G. INODE table entries	400	500	<u>600</u>
H. Other system tables	a	b	<u>c</u>

* The starting levels are indicated by an underscore.

The *file system distributions* (factor B) a, b and c refer to three specific algorithms used for distributing the user and system files among the disk drives. Obviously, the actual distribution depends on the number of disk drives used in a particular system configuration. Since the internal entropy (a measure of the lack of order in storing the files) could have a significant impact on response time, the team took care to preserve the internal entropy while changing from one file system to another during the experiments.

One system administrator suggested increasing the *memory size* (factor C) to improve the response time. However, another expert opinion was given that stated additional memory would not improve the response for the particular computer system being studied. Therefore, the team decided not to purchase more memory until they were reasonably sure its cost would be justified. They took level C_1 as the existing memory size, namely 4 MB, and disabled some of the existing memory to form levels C_2 and C_3. They decided to purchase more memory only if the experimental data showed that disabling a part of the memory leads to a significant reduction in performance.

Total memory is divided into two parts: *system buffers* (factor D) and user memory. The system buffers are used by the operating system to store recently used data in the hope that the data might be needed again soon. Increasing the size of the systems buffers improves the probability (technically called *hit ratio*) of finding the needed data in the memory. This can contribute to improved performance, but it also reduces the memory available to the users, which can lead to progressively worse system performance. Thus, the optimum size of the system buffers can depend on the particular load pattern. We refer to the size of the system buffers as a fraction of the total memory size. Thus, the levels of the system buffers are sliding with respect to the memory size.

Sticky bit, (factor E) is a way of telling an operating system to treat a command in a special way. When the sticky bit for a command such as *rm* or *ed* is set, the executable module for that command is copied contiguously in the swap area of a disk during system initialization. Every time that command is needed but not found in the memory, it is brought back from the swap area expeditiously in a single operation. However, if the sticky bit is not set and the command is not found in the memory, it must be brought back block by block from the file system. This adds to the execution time.

In this case study, factor E specifies how many and which commands had their sticky bits set. For level E_1, no command had its sticky bit set. For level E_2, the three commands that had their sticky bits set were *sh, ksh,* and *rm.* These were the three most frequently used commands during the month before the case study according to a 5-day accounting command summary report. For the level E_3, the eight commands that had their sticky bits set were the three commands mentioned above, plus the next five most commonly used commands, namely, the commands *ls, cp, expr, chmod,* and *sadc* (a local library command).

KMCs factor (factor F) are special devices used to assist the main CPU in handling the terminal and remote job entry traffic. They attempt to reduce the number of interrupts faced by the main CPU. In this case study, only the KMCs used for terminal traffic were changed. Those used for the remote job entry links were left alone. For level F_1, two KMCs were used to handle the terminal traffic, whereas for level F_2 the two KMCs were disabled.

The number of entries in the *INODE table* (factor G) determines the number of user files that can be handled simultaneously by the system. The three levels for the factor G are 400, 500, and 600. The three levels of the eighth factor, namely, the *other system tables* (factor H), are coded as a, b, and c.

Note that the software factors (B,D,E,G, and H) can affect only the response time. However, the three hardware factors (A,C, and F) can affect both the response time and the computer cost. Therefore, this optimization problem is not a pure parameter design problem, but, rather, a hybrid of parameter design and tolerance design.

DESIGN OF THE MATRIX EXPERIMENT AND THE EXPERIMENTAL PROCEDURE

This case study has seven 3-level factors and one 2-level factor. There are 7 x (3-1) + 1 x (2-1) + 1 = 16 degrees of freedom associated with these factors. The orthogonal array L_{18} is just right for this project because it has seven 3-level columns and one 2-level column to match the needs of the matrix experiment. The L_{18} array and the assignment of columns to factors are shown in Table 2. Aside from assigning the 2-level factor to the 2-level column, there is really no other reason for assigning a particular factor to a particular column. The factors were assigned to the columns in the order in which they were listed at the time the experiment was planned. Some aspects of the assignment of factors to columns are discussed next.

The experiment team found that changing the level of disk drives (factor A) was the most difficult among all the factors because it required an outside technician and took three to four hours. Consequently, in conducting these experiments, the team first conducted all experiments with level A_1 of the disk drives, then those with level A_2 of the disk drives, and finally those with level A_3 of the disk drives. The experiments with level A_3 of the disk drives were kept for last to allow time for the RP07 disk to arrive. However, because the RP07 disk did not arrive in time, the experimenters redefined level A_3 of the disk drives to be the same as level A_2 and continued with the rest of the plan. According to the dummy level technique discussed in Phadke, M. S. (1989), this redefinition of level does not destroy the orthogonality of the matrix experiment. This arrangement, however, gives 12 experiments with level A_2 of the disk drives; hence, more accurate information about that level is obtained when compared to level A_1 . This is exactly what we should look for because level A_2 is the new level which we have less prior information.

File distribution (factor B) was the second most difficult factor to change. Therefore, among the six experiments with level A_1 of the disk drives, the experiment team first conducted the two experiments with level B_1, then those with level B_2, and finally those with level B_3. The same pattern was repeated for level A_2 of the disk drives and then level A_3 of the disk drives. The examination of the L_{18} array given in Table 10.2 indicates that some of the bookkeeping of the experimental conditions could have been simplified if the team had assigned factor A to Column 2 and factor B tp Column 3. However, the inconvenience of the particular column assignment was unimportant.

For each experiment corresponding to each row of the L_{18} array, the team ran the system for two days and collected data on standard response time and trivial response time once every ten minutes from 9:00 a.m. to 5:00 p.m. While measuring the response times, they made sure that the UNIX system *cron* facility did not schedule some routine data collection operations at the same exact moments because those would affect the measurement of response times.

Running experiments on a live system can invite a number of practical problems. Therefore, the first thing the experimenters did was to seek the cooperation of users. One problem of great concern to

Table 2 Orthogonal Array and Factor Assignment

Expt #	Column # and Factor Assignment							
	1 F	2 B	3 C	4 D	5 E	6 A	7 G	8 H
1	1	1	1	1	1	1	1	1
2	1	1	2	2	2	2	2	2
3	1	1	3	3	3	3	3	3
4	1	2	1	1	2	2	3	3
5	1	2	2	2	3	3	1	1
6	1	2	3	3	1	1	2	2
7	1	3	1	2	1	3	2	3
8	1	3	2	3	2	1	3	1
9	1	3	3	1	3	2	1	2
10	2	1	1	3	3	2	2	1
11	2	1	2	1	1	3	3	2
12	2	1	3	2	2	1	1	3
13	2	2	1	2	3	1	3	2
14	2	2	2	3	1	2	1	3
15	2	2	3	1	2	3	2	1
16	2	3	1	3	2	3	1	2
17	2	3	2	1	3	1	2	3
18	2	3	3	2	1	2	3	1

the users was that for a particular combination of control factor settings, the system performance could become bad enough to cause a major disruption in their activities. To avoid this, the system administrator was instructed to make note of such an event and go back to the pre-experiment settings of the various factors. This would minimize the inconvenience to the users. Fortunately, such an event did not occur, but had it happened, the experiment team would still have been able to analyze the data and determine optimum conditions using the accumulation analysis method described in Phadke, M. S. (1989) (see also Taguchi, G. (1977), and Taguchi, G. and Wu, Yn-In (1979)).

Under the best circumstances, the team could finish two experiments per week. For 18 experiments, it would take nine weeks. However, during the experiments, a snowstorm arrived and the Easter holiday was observed, both events causing the system load to drop to an exceptionally low level. Those days were eliminated from the data and the team repeated the corresponding combinations of control factor settings to have data for every row of the matrix experiment.

DATA ANALYSIS AND VERIFICATION EXPERIMENTS

From the 96 measurements of standard response time for each experiment, the team computed the mean response time and the S/N ratio. The results are shown in Table 3. Similar computations were made for the trivial response time, but they are not shown here. The effects of the various factors on the S/N ratio for the standard response time are shown, along with the corresponding ANOVA in Table 4. The factors effects are plotted in Figure 3. Note that the levels C_1, C_2, and C_3 of memory size are 4.0 MB, and 3.5 MB, respectively, which are not in a monotonic order. While plotting the data in Figure 3, the experimenters considered the correct order.

It is apparent from Table 4 that the factor effects are rather small, especially when compared to the error variance. The problem of getting large error variance is more likely with live experiments, such as this computer system optimization experiment, because the different rows of the matrix experiment are apt to see quite different noise conditions, that is, quite different load conditions. Also, while running live experiments, the tendency is to choose the levels of control factors that are not far apart.

However, we can still make valuable conclusions about optimum settings of control factors and then see if the improvements observed during the verification experiment are significant or not.

The following observations can be made from the plots in Figure 3 and Table 4 (note that these conclusions are valid only for the particular load characteristics of the computer being tuned):

1. Going from not setting any sticky bits to setting sticky bits on the three most used commands does not improve the response time. This is probably because those three commands tend to stay in the memory as a result of their very frequent use, regardless of setting *sticky bits*. However, when *sticky bits* are set on the five next most used commands, the response time improves by 1.69 dB. This suggests that we should set *sticky bits* on the eight commands, and in future experiments, we should consider even more commands for setting *sticky bits*.

2. KMCs do not help in improving response time for this type of computer environment. Therefore, they may be dropped as far as terminal handling is concerned, thus reducing the cost of the hardware.

3. Adding one more disk drive leads to better response time. Perhaps even more disks should be considered for improving the response time. Of course, this would mean more cost, so proper trade-offs would have to be made.

4. The S/N ratio is virtually the same for 4 MB and 3.5 MB memory. It is significantly lower for 3 MB memory. Thus, 4 MB seems to be an optimum value-that is, buying more memory would probably not help much in improvement response time.

5. There is some potential advantage (0.8 dB) in changing the fraction of system buffers from 1/3 to 1/4.

6. The effects of the remaining three control factors are very small and there is no advantage in changing their levels.

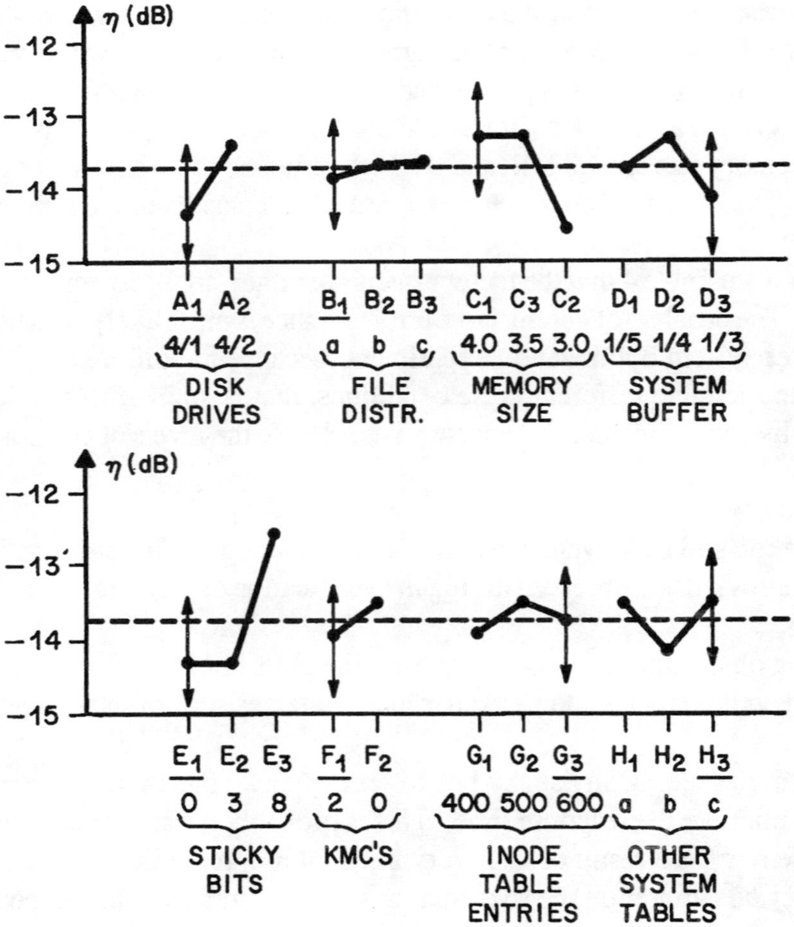

Figure 3 Factors for S/N Ratio for Standard Response
(Underscore indicates starting level. One -standard-deviation limits are shown)

Table 3 Data Summary for Standard Response Time

Expt. #	Mean (sec)	η (dB)
1	4.64	-14.66
2	5.28	-16.37
3	3.06	-10.49
4	4.53	-14.85
5	3.26	-10.94
6	4.55	-14.96
7	3.37	-11.77
8	5.62	-16.72
9	4.87	-14.67
10	4.13	-13.52
11	4.08	-13.79
12	4.45	-14.19
13	3.81	-12.89
14	5.87	-16.75
15	3.42	-11.65
16	3.66	-12.23
17	3.92	-12.81
18	4.42	-13.71

Table 4 Analysis of S/N Ratios for Standard Response*

Factor	Average η by Factor Level			degrees of freedom	sum of squares	mean squares	F
	1	2	3				
A. Disk Drives	-14.37	-13.40		1	3.76	3.76	1
B. File Distrib.	-13.84	-13.67	-13.65	2	0.12•	0.06	
C. Memory Size	-13.32	-14.56	-13.28	2	6.40	3.20	
D. System Buffers	-13.74	-13.31	-14.11	2	1.92	0.97	
E. Sticky Bits	-14.27	-14.34	-12.55	2	12.27	6.14	2
F. KMCs Used	-13.94	-13.50		1	0.84•	0.42	
G. INODE entries	-13.91	-13.51	-13.74	2	0.47•	0.24	
H. Other sys. tables	-13.53	-14.15	-13.48	2	1.68	1.34	
Error				3	32.60	10.87	
Total				17	60.06	3.53	
(Error)				(9)	(34.03)	(3.78)	

* Overall mean =-13.72 dB; underscore indicates starting conditions.

• Indicates the sum of squares that were added to form the pooled error sum of squares shown in parenthesis.

OPTIMUM CONDITIONS AND VERIFICATIONS EXPERIMENTS

The optimum system configuration inferred from the data analysis above is shown in Table 5 along with the starting configuration. Changes were recommended in the settings of sticky bits, disk drives, and system buffers because they lead to faster response. KMCs were dropped because they did not help improve response, and dropping them meant saving hardware. The prediction of the S/N ratio for the standard response time under the starting and optimum conditions is also shown in Table 5. Note that the contributions of the factors, whose sum of squares were among the smallest and were pooled, are ignored in predicting the S/N ratio. Thus, the S/N ratio predicted by the data analysis under the starting condition is -14.67 dB, and under the optimum conditions it is -11.22 dB. The corresponding predicted rms response times under the starting and optimum conditions are 5.41 seconds and 3.63 seconds, respectively.

Table 5 Optimum Settings and Prediction of Standard Response Time

	Starting Condition		Optimum Condition	
Factor	**Setting**	Contribution• (dB)	**Setting**	Contribution• (dB)
A. Disk Drives*	A_1	-.65	A_2	0.32
B. File Distribution	B_1	-	B_1	-
C. Memory Size	C_1	0.40	C_1	0.40
D. System Buffers*	D_3	-.39	D_2	0.41
E. Sticky bits*	E_1	0.55	E_3	1.17
F. KMCs used*	F_1	-	F_2	-
G. INODE table entries	G_3	-	G_3	-
H. Other system tables	H_3	0.24	H_3	0.24
Overall mean		-13.72		-13.72
Total		-14.67		-11.22

As noted earlier, here the error variance is large. We would also expect the variance of the prediction error to be large. The variance of the prediction error can be computed by the procedure given in Chapter 3 [see Equation (3.14)]. The equivalent sample size for the starting condition n_e, is given by

$$\frac{1}{n_e}=\frac{1}{n}+\left(\frac{1}{n_{A_1}}-\frac{1}{n}\right)+\left(\frac{1}{n_{C_1}}-\frac{1}{n}\right)+\left(\frac{1}{n_{D_3}}-\frac{1}{n}\right)+\left(\frac{1}{n_{E_1}}-\frac{1}{n}\right)+\left(\frac{1}{n_{H_1}}-\frac{1}{n}\right)=\frac{1}{18}$$

$$+\left(\frac{1}{6}-\frac{1}{18}\right)+\left(\frac{1}{6}-\frac{1}{18}\right)+\left(\frac{1}{6}-\frac{1}{18}\right)+\left(\frac{1}{6}-\frac{1}{18}\right)+\left(\frac{1}{6}-\frac{1}{18}\right)=0.61.$$

Thus, n_e =1.64. Correspondingly, the two-standard-deviation confidence limits for the predicted S/N ratio under the starting conditions are $-14.67\pm 2\sqrt{3.78/1.64}$, which simplify to $-14.67\pm 3.04\text{dB}$. Similarly, the two-standard-deviation confidence limits for the predicted S/N ratio under the optimum conditions are $-11.22\pm 2\sqrt{3.78/1.89}$, which simplify to $-11.22\pm 2.83\text{dB}$. Note that there is a slight difference between the widths of the two confidence intervals. This is because there are 12 experiments with level A_2 and only six with A_1 .

Subsequently, the optimum configuration was implemented. The average response times for the standard and trivial commands for over two weeks of operation under the optimum conditions are plotted in Figure 1(a)-(b). Comparing the response time under the starting (initial) configuration with that under the optimum configuration, we see that under the optimum conditions the response time is small, even in the afternoon when the load is high. In fact, the response time is uniformly low throughout the day.

The data from the confirmation experiment are summarized in Table 6. For standard response time we see that the mean response time was reduced from 6.15 sec.to 2.37 sec., which amounts to a 61-percent improvement. Similar improvement was seen in the rms response time. On the S/N ratio scale, the improvement was 8.39 dB. Similarly, for the trivial response time, the improvement was seen to be between 70 percent and 80 percent of the mean value, or 13.64 dB, as indicated in Table 6.

Table 6 Results of Verification Experiment

Measure	**Starting Levels**	Optimum Levels	Improve-ment	**Starting Levels**	Optimum Levels	Improve-ment
Mean (sec)	6.15	2.37	61%	0.521	0.148	71%
rms (sec)	7.59	2.98	61%	0.969	0.200	79%
η (dB)	-17.60	-9.21	8.39dB	+0.34	+13.98	13.64 dB
Standard-ized η (dB)	-15.88	-8.64	7.24dB	+3.26	+15.71	12.45 dB

Note that the observed S/N ratios for standard response time under the starting and optimum conditions are within their respective two-standard-deviation confidence limits. However, they are rather close to the limits. Also, the observed improvement (8.39 dB) in the S/N ratio is quite large compared to the improvement predicted by the data (3.47 dB). However, the difference is well within the confidence limits. Also, observe that the S/N ratio under the optimum conditions is better than the best among the 18 experiments.

Thus, here we achieved a 60- to 70-percent improvement in response time by improved system administration. Following this experiment, two similar experiments were performed by Klinger, W. J. and Nazaret, W. A.(1985) who took extra care to ensure that published UNIX system tuning guides were used to establish the starting conditions. Their experiments still led to a 20- to 40-percent improvement in response time. One extra factor they considered was the use of PDQs, which are special auxiliary processors for handling text processing jobs (troff). For their systems, it turned out that the use of PDQs could hurt the response time rather than help.

STANDARDIZED S/N RATIO

When running Robust Design experiments with live systems, we face the methodological problem that the noise conditions, which are the load conditions for our computer system optimization, are not the same for every row of the control orthogonal array. This can lead to inaccuracies in the conclusions. One way to minimize the impact of changing noise conditions is to construct a standardized S/N ratio, which we describe next.

As noted earlier, some of the more important load measures for the computer system optimization experiment are: number of users, CPU demand, I/O demand, and memory demand. After studying the load pattern over the days when the case study was conducted, we can define low and high levels for each of these load measures, as shown in Figure 4. These levels should be defined so that for every experiment we have a reasonable number of observations at each level. The 16 different possible combinations of the levels of these four load measures are listed in Table 7 and are nothing more than 16 different noise conditions.

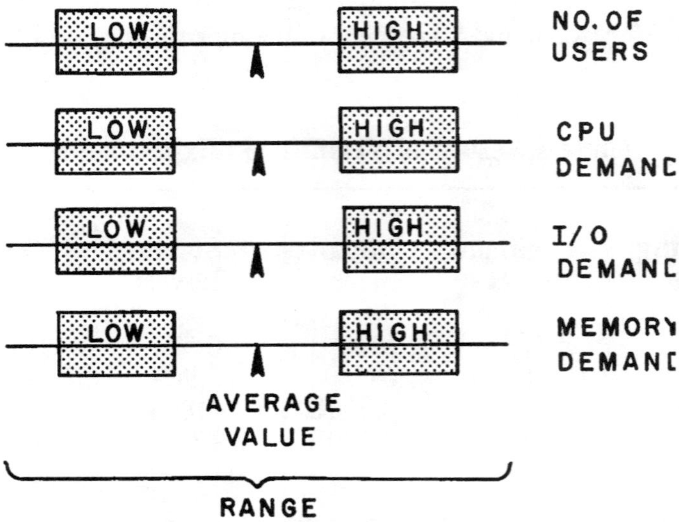

Figure 4 Selection of Levels for Load Measures

In a live experiment, the number of observations for each noise condition can change from experiment to experiment. For instance, one day the load might be heavy while another day it might be light. Although we cannot dictate the load condition for the different experiments, we can observe the offered load. The impact of load variation can be minimized as follows: We first compute the average response time for each experiment in each of the 16 load conditions. We then treat these 16 averages as raw data to compute the S/N ratio for each experiment. The S/N ratio computed in this manner is called standardized S/N ratio because it effectively standardizes the load conditions for each experiment. The system can then be optimized using this standardized S/N ratio.

Note that for the standardized S/N ratio to work, we must have good definitions of noise factors and ways of measuring them. Also, in each experiment every noise condition must occur at least once to be able to compute the average. In practice, however, if one or two conditions are missing, we may compute the S/N ratio with the available noise conditions without much harm.

In the computer system optimization case study discussed in this paper, the experimenters used the concept of standardized S/N ratio to obtain better comparison the starting and optimum conditions. Some researchers expressed a concern that part of the improvement observed in this case study might have been due to changed load conditions over the period of three months that it took the team to conduct the matrix experiment. Accordingly, the experimenters computed the standardized S/N ratio for the two conditions and the results are shown in Table 6 along with the other results of the verification experiment. Since the improvement in the standardized S/N ratio is quite close to the original S/N ratio, the experiment team concluded that the load change had minimal impact on the improvement.

RELATED APPLICATIONS

Computer System Parameter Space Mapping

In practice, running live experiments to optimize every computer installation is obviously not practicable. Instead, more benefits can be achieved by performing off-line matrix experiments to optimize for several different load conditions. Such load conditions can be simulated using facilities like the UNIX system benchmarking facility. The information gained from these off-line experiments can be used to map the operating system parameter space. This catalog of experimental results can then be used to improve the performance of different machines.

Optimizing Operations Support Systems

Large systems, such as the traffic in a telecommunications network, are often managed by software systems that are commonly known as operations support systems. The load offered to a telephone network varies widely from weekday to weekend, from morning to evening, from working days to holidays, etc. Managing the network includes defining strategies for providing the needed

Table 7 Computation of Standardized S/N Ratio*

Expt. #	# of Users	CPU demand	I/O demand	Memory Demand	Average Re - sponse Time for Expt. i
1	1	1	1	1	yi1
2	1	1	1	2	yi2
3	1	1	2	1	yi3
4	1	1	2	2	yi4
5	1	2	1	1	yi5
6	1	2	1	2	yi6
7	1	2	2	1	yi7
8	1	2	2	2	yi8
9	2	1	1	1	yi9
10	2	1	1	2	yi10
11	2	1	2	1	yi11
12	2	1	2	2	yi12
13	2	2	1	1	yi13
14	2	2	1	2	yi14
15	2	2	2	1	yi15
16	2	2	2	2	

*Standardized S/N ratio for experiment i,

$$\eta_1 = -10\log_{10}\left(\frac{1}{16}\sum_{j=1}^{16} y^2_{ij}\right)$$

transmissions and switching facilities, routing a call end to end on the network, and many other tasks. Steve Eick and Madhav Phadke applied the Robust Design method in AT&T to improve the strategy for adapting the routing algorithm to handle unusually high demands by as much as 70 percent. This not only improved the revenue, but also improved network performance by reducing the number of blocked calls. These experiments were done on a simulator because of the high risks of the live experiments.

International telephone traffic is carried over several different types of transmission facilities. Seshadri, V.(1988) used the Robust Design method to determine the preferred facility for transmitting facsimile data by conducting live experiments on the network.

The success in improving the telephone network traffic management suggests that the Robust Design approach could be used successfully to improve other networks, for example, air traffic management. In fact, it is reported that the method has been used in Japan to optimize runway and air-space usage at major airports.

REFERENCES

Eick, S. G. and Phadke, M. S.," Improving the performance of network management reroute algorithm using the Robust Design method", included in the tutorial titled Total Quality Management in Telecommunications, *IEEE Globeecom 90*, San Diego, CA, December 1990.

Klinger, W. J. and Nazaret, W. A.(1985), " Tuning Computer Systems for Maximum Performance: A Statistical Approach." Computer Science and Statistics: Proceedings of the 18th Symposium of the Interface, Fort Collins, Colorado, March 1985, pp. 390-396.

Pao, T. W., Sherrerd, C. S , and Phadke, M. S. (1985), "Computer Response Time Optimization Using Orthogonal Array Experiments", IEEE International Communications Conference, Chicago, IL, June 23-26, 1985, vol. 2, pp. 890-895.

Phadke, M. S. (1989), 'Quality Engineering using Robust Design', Prentice Hall, 1989.

Seshadri, V. (1988), " Application of the Taguchi Method for Facsimile Performance Characterization on AT&T International Network." Proceedings of IEEE International Conference on Communications, Philadelphia, PA, June 1988, pp. 0833-0835.

Taguchi, G. (1977), 'Jikken Keikakuho', 3rd Edition, Tokyo, Japan: Maruzen, vol. 1 and 2, 1977 and 1978 (in Japanese). English translation: Taguchi, G. System of Experimental Design. Edited by Don Clausing. New York: UNIPUB/Kraus International Publications, vol. 1 and 2, 1987.

Taguchi, G. and Wu , Yn-In (1979), "Introduction to Off-Line Quality Control". Central Japan Quality Control Association. Meieki Nakamura-Ku Magaya, Japan, 1979. (Available from American Supplier Institute, Inc., Dearborn, MI).

---

* VAX is a registered trademark of Digital Equipment Corporation (Page 98).

QUALITY DESIGN ENGINEERING

(The Missing Link in U.S. Competitiveness)

H. Barry Bebb*

SUMMARY

"You criticize us all the time for not buying American products while it is obvious that you prefer Japanese products. Are you asking us to buy something you won't buy yourself?"
Shintaro Ishihara, The Japan That Can Say "No" Simon & Shutzer, 1989
(Quotes taken from earlier version which was translated from Japanese; it was co-authored by Akio Morita)

Observations such as this have prompted general discussions in U. S. corporate boardrooms about the importance of product development issues. These discussions often focus on topics such as Technology, Business Strategy, Product Development Process and Manufacturing Quality. However, they seldom focus on a root-cause of U.S. product development problems, namely, the **quality of design engineering** of mechanical, electro-mechanical and electro-opto-mechanical systems found in most products in the home, office or factory and on highways, water or in the air. In this paper, Competitive Benchmarking by R. C. Camp (Camp, 1989) and earlier work by John R. Dixon and Michael R. Duffey (Dixon and Duffey, 1990) are evoked to identify high quality design engineering as the "missing link" in efforts to achieve and maintain global competitiveness. Based on the author's experience at Xerox, a new model is proposed for organizing, managing, and communicating design engineering practices. Also, specific recommendations are formulated in terms of "change agents" for continuous process improvement.

* The author, currently with Barry Bebb & Associates, was formerly with Xerox Corporation where the work presented in this paper was performed.

THE ROLE OF DESIGN ENGINEERING

Design engineering is the set of processes that translates customer requirements into manufacturable outcomes. It involves the detailed design of parts and assemblies, their integration into an engineering prototype model, and finally an operational representation of the final, manufacturable product. Design engineering requires a rich repertoire of technical disciplines and craft skills such as geometry creation (drawings), generation of bills of material, materials selection, tolerance analyses, stress/strain analyses, and performance latitude analyses. Design-for-manufacturability, and knowledge of fabrication and assembly of parts is also crucial for modern concurrent engineering processes.

It is in *mechanical* design engineering where our greatest competitive shortfalls occur. Our design engineering problems in electronics and software are less severe, and they do not impact quality, cost and schedule outcomes for products like copiers, appliances, automobiles, and video equipment to the same extent as the shortfalls in mechanical elements. However, experience shows that the subject of mechanical design is not popular - it does not have the magnetic attraction of an exciting new technology breakthrough that causes senior management to clamor to get involved.

EVIDENCE FOR THE "MISSING LINK"

Evidence for mechanical design engineering as a crucial "missing link" has increasingly surfaced at Xerox and other companies in the last decade. In competitive benchmarking studies performed at Xerox in 1979 and 1980, manufacturing cost and quality were compared in Japanese and American parts and assemblies. The importance of design engineering was identified in two fundamental ways:

First, benchmarking of manufacturing outcomes led to the judgment that *more than 50% of cost and quality of the manufactured product is directly attributable to design.* One study revealed a 50% cost gap between plastic parts from American, European, and Latin American suppliers compared with Japanese suppliers. The cost gap was segmented into resin costs, the costs of processing and tools, and the cost attributed to design engineering. Each of the segments was further partitioned as shown in Figure 1. This analysis revealed that resins could be purchased in Japan for

Figure 1: Breakdown of the 50% gap found during competitive benchmarking against Japanese copier manufacturers

Resign: 10%	**Processing & Tools: 15%**	**Design Engineering: 25%**
	Mold cost	Tolerances
	Mold preparation	Multi-functionality
	Mold change time	Wall thickness
	# of injections for first good part	Inserts
	Cycle time	Tooling design
	Material handling	Finishing
	Packaging	
	Labor	

10% less than in the western world; process and tooling was 15% less; and the remaining 25% was attributable to how the parts were designed. The Japanese were simply out-designing us: more functions per plastic part, fewer inserts, better manufacturability of product designs and associated mold designs, fewer finishing operations, and so forth.

This relationship between design engineering and manufacturing outcomes was also clearly seen by comparing parts and assemblies manufactured at multiple sites but designed at one engineering site (this is particularly pertinent for multinational corporations like Xerox with manufacturing sites in multiple countries around the world). Parts and assemblies that are well designed in one country are readily produced as high quality and low cost parts and assemblies in other countries, including third world countries; bad designs are not.

The second and most revealing result from benchmarking studies indicates that *design quality achieved prior to the first engineering prototype-build largely determines how many prototype cycles will be required and the total time-to-market of the product.* Figure 2 shows a variation of L. P. Sullivan's (Sullivan, 1986) well known graph comparing engineering change rates between the U.S. and Japan. A Japanese development team will typically design, build, test, and refine a single set of prototype units (P) and then proceed to the first manufacturing block (B). A comparable American team may require, say, three full prototype design, build, test, and retrofit cycles (P1, P2, and P3) with a total of four times the number of engineering changes spread over a longer development cycle. Using some admittedly gross assumptions, the axis labeled "Number of Engineering Changes Processes" can also be labeled "Number of Engineers".

Typical criteria that enable a prototype to serve as the basis for the manufacturing build (B) might be a design quality that (1) yields prototype models that operate with failure rates no greater than a factor of ten higher than final product specifications; and (2) requires that no more than 50% of the drawings need changes prior to the B build. The comparisons in Figure 2 show that American teams reach this level of design quality at P3; Japanese teams reach it at P1.

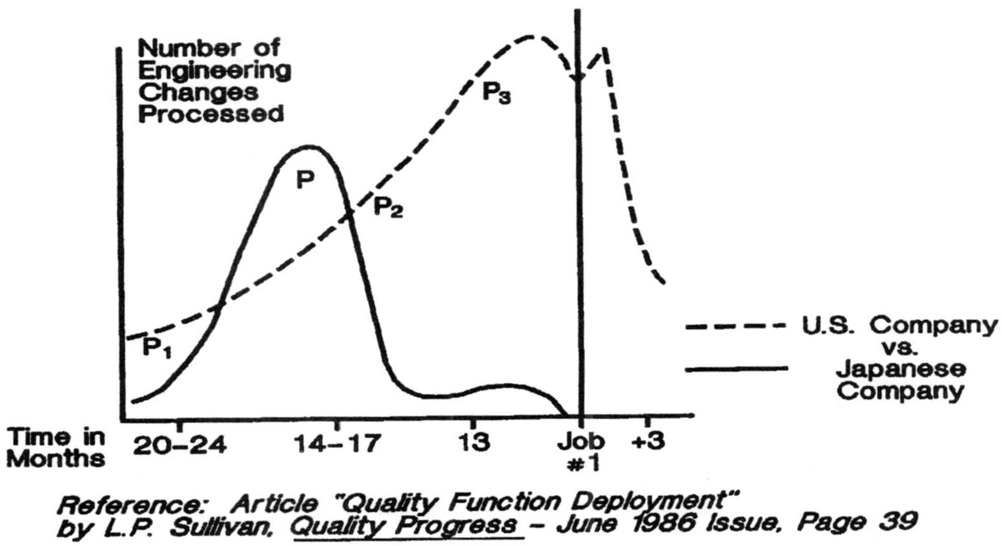

Figure 2 Benchmarking Comparison of Japanese vs. U. S. Engineering Changes

A MODEL FOR DESIGN ENGINEERING PROCESSES

A very useful vehicle for organizing and communicating design engineering concepts is the "Bottle" model in Figure 3. The center axis of the "Bottle" curves is the design set point (a specified set of operating parameters with which to reference design performance). The 'Bottle" curve that opens up from left to right over time depicts the product design progressing toward *increasing latitudes of operation* (robustness). Latitude means the range of possible operating conditions in which the product can perform satisfactorily. In general, a product is first designed to meet nominal, target performance specifications; then the design latitude is increased as the design proceeds. The "Bottle" that closes down from left to right depicts the manufacturing process design dynamics of reducing variances (noises) between multiple units of the same design. This designed-in control of manufacturing build to on-going production. Design quality is defined graphically in the "Bottle" model as the difference between the operational latitude curves and the manufacturing variance curves around the design set point.

The "Bottle" depiction has been ten years in the making. Don P. Clausing (Clausing, 1980) crafted the first version of this chart to portray the notion of increasing latitudes and decreasing variances as the design proceeds. In about 1985, Maurice Holmes (1985) integrated the Taguchi (Baker and Clausing, 1984; Taguchi; 1987; Byrne and Taguchi, 1987) methodology related to "robustness against noise" and "noises" to identify the difference between latitudes and variances as Design Quality.

This author made several additions to the emerging model:

1. Choose the "Bottle" shapes as appropriate representations of the iterative engineering design processes of simultaneously increasing latitudes and reducing variances,
2. Identified the difference between latitude and variance with the Taguchi Quality Loss Function as a measure of Design Quality,
3. Related specific engineering design practices to the generic phases of product development, and
4. Related the "Bottle" parameters to the various upper and lower limits referenced in Statistical Process Control (SPC) methodologies.

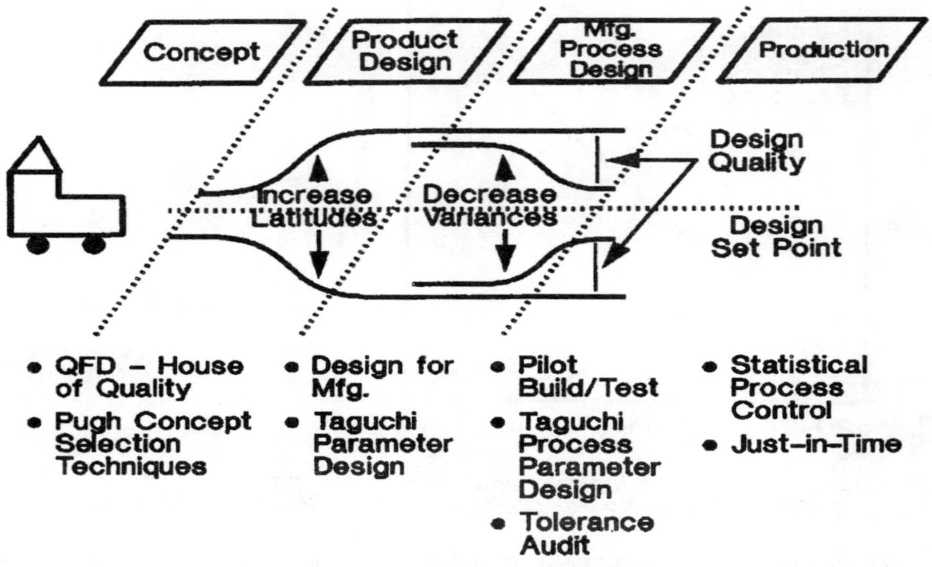

Figure 3: The" Bottle" Model and Design Quality

The "Bottle"" construct was first introduced in the Fall of 1987 in a presentation by the author to a meeting of the National Academy of Engineering about U.S. competitiveness in the fall of 1987 (Bebb, 1987). The model has proved to be valuable for organizing, managing, and communicating the engineering design practices in the context of the broader product development processes.

THE "BOTTLE" MODEL AND CONCURRENT ENGINEERING DESIGN

Good engineering design requires early attention to manufacturing requirements, and good manufacturing design requires continuous attention to engineering design requirements. The necessary integration of engineering and manufacturing teams to break down traditional interface barriers has recently been popularized by the descriptive term *concurrent engineering design*. The "Bottle" model depicts the concurrent engineering philosophy through the overlap of the latitude and variance bottles. For example, the "Bottle" associated with Manufacturing Process Design to depict the compressed Japanese design schedules that involve a single-cycle process for design, build, test, and fix of a prototype prior to the first manufacturing block (B) design-and-build cycle. Conversely, the "Bottles" can be further separated along the time axis to select longer schedules involving multiple prototypes that typify American design processes. Other variations are possible such as changing the slope of the transition curves to reflect different rates of improving design quality and reducing variances or replacing the smooth curves with "stair steps" to emphasize the iterative character of the design process. The dotted lines in Figure 3 that separate the phases are off-vertical to suggest that some overlap between the phases is appropriate to the iterative design processes.

The notion of Concurrent Engineering Design must be extended beyond the integration of engineering and manufacturing. The broader sets of disciplines include Customers, Marketing, Product Planning, Engineering, Manufacturing (including suppliers), Service, Human Factors, Industrial Design, Safety, Environmental Engineering, Distribution, Marketing, and Sales. In many product development environments, **service** and **customer** representatives are the most critical additions to a team. If, for example, service is not fully represented during the early portions of detailed design but achieves full staffing later in the program, then the serviceability assumptions made by engineering and manufacturing team members will almost surely be subject to change. Such changes result in Engineering Change notices (reference Figure 2), and excessive changes result in the necessity of another complete Engineering Prototype build cycle with the associated negative impacts on quality and schedule.

To further emphasize the criticality of adequately staffing all disciplines throughout the development process, consider the conflict in processes associated with Design for Automated Assembly and Design for Serviceability. Good Design for Assembly normally yields sequential assembly processes. Good Design for Serviceability strives to facilitate access to a failed part with a minimum number of actions rather than sequentially removing a large number of parts. If engineering and manufacturing have joined to optimize Design for Assembly without strong service participation, serviceability will either be unsatisfactory or require redesign. Similar scenarios apply to the other functions. In this context, **it is useful to regard service as re-manufacturing in field environments that are more hostile than new build factory conditions.**

THE "BOTTLE" MODEL AND PROTOTYPE BUILD CYCLES

A benchmark single-cycle process for design, build, test, and fix of a prototype is compared with a three-prototype-cycle process in Figures 4a and 4b. At the start of detailed design, both examples show variances exceeding latitudes. In the benchmark single-prototype-cycle, latitudes exceed variances when the first design is completed; each prototype unit should then operate over the full specified range of controllable parameters (say, engine throttle position and ignition timing) and uncontrollable parameters (say, temperature and humidity), they perform well around nominal parameter values. For this benchmark single-cycle process, subsequent fine-tuning of the prototype should produce a functionally robust and manufacturable design.

Multiple prototype cycles, however, are more typical of American design engineering processes. When the first design is completed for the three-prototype-cycle process in Figure 4b, variances still exceed latitudes. Different P1 prototype units would probably require different set-up parameters, and considerable tinkering and adjusting would be necessary just to make the models run. Clearly a design that yields unit-to-unit variation is not sufficiently robust for quantity

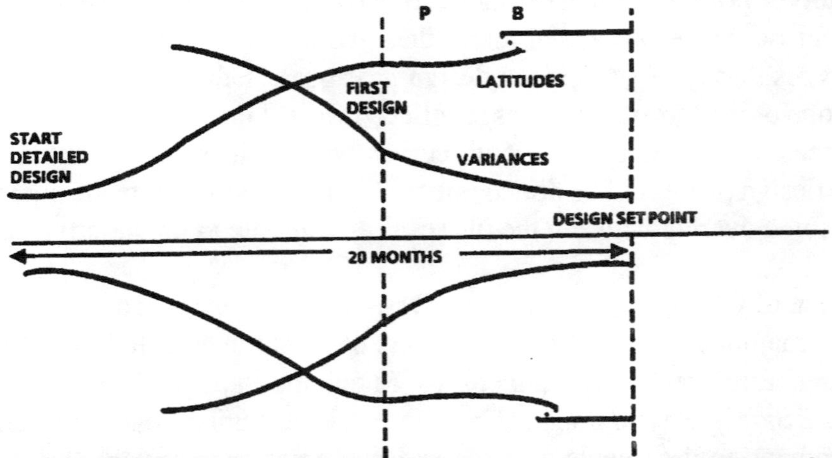

Figure 4a: Single Prototype Design, Build, Test and Fix Cycle: Typical of Japanese Manufacturers

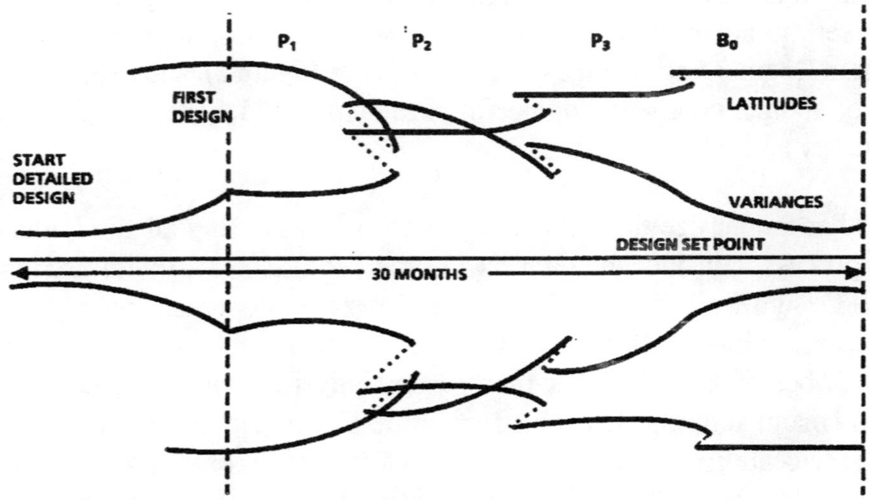

Figure 4b: Three Prototype Design, Build, Test, and Fix Cycles: Typical of U. S. Manufacturers

production. The subsequent prototype is then designed with increased operational latitudes. However, variances typically deteriorate between the end of one prototype cycle and the beginning of the next one since the new units must be debugged, tuned up, and exercised. These transitions are depicted in the "Bottle" model as breaks connected with dotted lines (this type of overlapped schedule is sometimes called the Z-curve schedule).

The character of the curves representing the three-prototype-cycle reflects a very real chaos. When a prototype build commences, most of the design team becomes consumed with the tasks of assembling and debugging the prototypes - chasing parts, assembling things, debugging things, and so forth. The *urgent* wins priority over the *important*. The grander notions of increasing latitudes and reducing variances succumb to the activities necessary to get the right parts with the right holes in the right places. Once the build, test, and fix cycles start, design becomes more directed towards finding and fixing problems than making improvements in latitudes and variances, both within and between prototype cycles.

Perhaps a worse problem in multiple prototype cycles relates to the team member expectations. Team members can and will meet visible and inspectable schedule checkpoints by delaying designs until the next cycle (design quality and completeness are difficult to inspect). Incomplete designs preclude total test coverage. Hence, problem identification and correction will be delayed. Other complexities could be cited; for brevity, the discussion will be terminated with the observation that *if three prototype cycles are planned, it is certain that no fewer than three cycles will be executed.*

Experience and research show that bad first designs are irrecoverable (Harry, 1989). Even with increased schedule time to accommodate multiple prototype build cycles, the quality of the last prototype design and the quality of the manufactured product are plagued with residual quality problems from the first prototype design. The benchmark single prototype cycle process does not contain the same complexities; it provides for increased quality time to pursue the grander notions of design quality; it creates team expectations to achieve a quality design and a complete design prior to the first prototype cycle; and it reduces the chaos generated by multiple, overlapping prototype cycles. By simply establishing high quality design objectives for the first prototype build, an attitude is set for the product development team that will almost certainly enhance the quality of first prototype models.

THE "BOTTLE" MODEL AND EVALUATION OF DESIGN QUALITY

As previously mentioned, *design quality* is defined in the "Bottle" diagram as the difference between operational latitude curves and manufacturing variance curves. However, this graphic definition belies the complexity of procedures necessary for the design engineer to calculate the bottle parameters. The variance and latitude curves are actually extreme of distributions around the design set point. They are determined by varying the critical parameters to define extreme for some chosen distribution (e.g., the traditional three sigma point for extreme of normal distributions).

Ideally, design quality should be evaluated prior to the first prototype build by use of simulation and designed experiments coupled with test beds of selected subsystems and modules. Subsequent evaluation of the prototype units should then validate this pre-build design quality. In practice, however, techniques for evaluating pre-build design quality are often not available (or simply outside the experience of the design team). Evaluation of the completed prototype is instead used to correct design shortfalls in latitudes and variances for a subsequent prototype build.

The methodologies pioneered by Genichi Taguchi (Baker and Clausing, 1984; Taguchi; 1987; Byrne and Taguchi, 1987) have proven very effective for both characterizing the "Bottle" parameters and evaluating design quality. The "Bottle" depiction of opening latitudes and closing variances relates directly to Taguchi's orthogonal array methodologies. Closing of variances (noise) can be accomplished through application of Taguchi Manufacturing Process Parameter Design. At any time in the design process, the design quality can be determined by relating the Taguchi Quality Loss Function (Barker and Clausing, 1984; Taguchi; 1987; Byrne and Taguchi, 1987) to the differences in latitudes and variances. Taguchi Parameter and Tolerance Design processes can also be used for improving Design Quality. Also, design quality early in the design engineering process is directly related to *manufacturing quality* as measured by Statistical Process Control (SPC) techniques during the production process. The control and distribution charts popularized by Deming for SPC relate directly to the "Bottle" model as illustrated in Figure 5. The upper and lower *control limits* are set at product design latitudes. The upper and lower *specification limits* tie to manufacturing process design variances. The jagged line represents reductions in variances that are normally achieved as a manufacturing process matures.

The distribution charts in Figure 5 depict the same improvements in variances between manufacturing start-up and stable-process control. The linking of SPC parameters and the bottle representation clarifies the interconnection between upstream design processes and downstream manufacturing processes in a way that brings to mind the adage that high and low cost cannot be sprinkled on in manufacturing and test.

Finally, system level bottle models can be constructed hierarchically from bottle models appropriate to parts which integrate into modules which in turn integrate into subsystems which in turn integrate into products. Tolerance and variance stack-up methodologies provide close analogies to the hierarchical construct for "Bottles". System level latitudes and variances are determined by

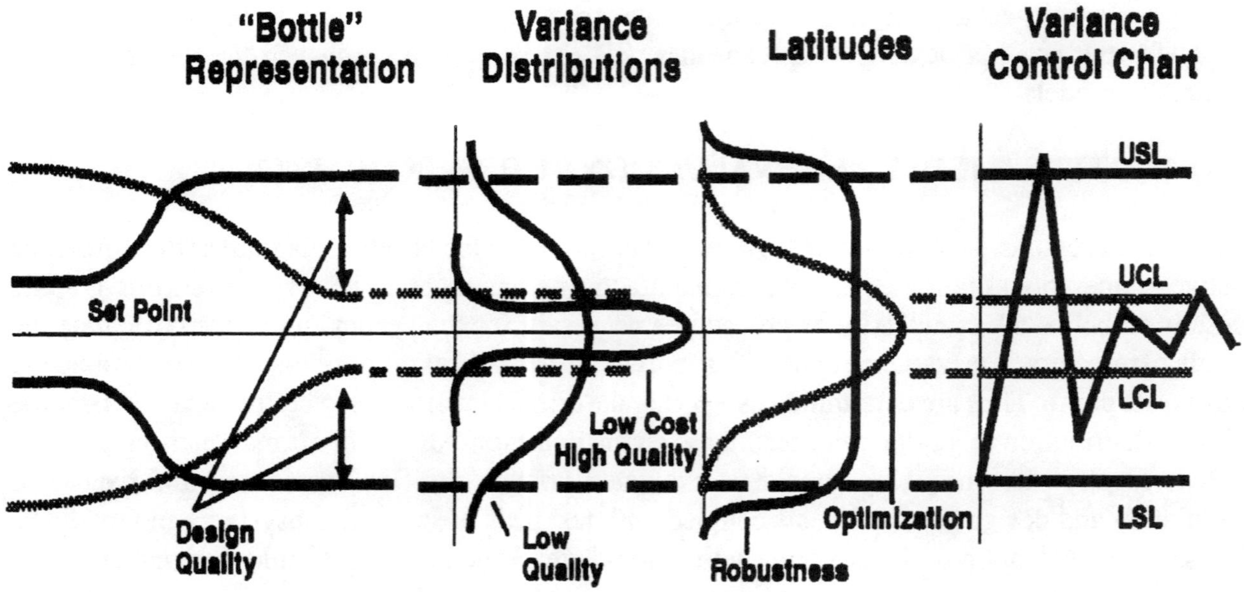

Figure 5: Relationship of " Bottle" with common Quality Measures

the **Robustness against Noise and the Controllable and Uncontrollable Noise characteristics** of the collection of elements that constitute the system (Taguchi, 1987). Analysis techniques for calculating stack-ups include C_{pk} examinations in the context of Motorola's 6-sigma perspectives (Harry, 1989) and focusing on the rates of reducing bugs or mistakes during the design process. Digital Equipment Corporation eliminated all but four percent of the bugs or design mistakes prior to the first hardware build of the VAX 8800 (Clark, 1990). **A second prototype-build was not required!**

THE "BOTTLE" MODEL AND CONTINUOUS QUALITY FUNCTION DEPLOYMENT

Good design practice starts with the customer. The first step is to synthesize often diffused customer requirements. The second step is to translate the customer requirements into operational product specifications. Design engineers must integrate not only the direct requirements of the *external paying customer* but the *internal customer requirements* of people in marketing, engineering, manufacturing, distribution, sales and service. Marketing people seek features, performance and price that yield competitive privileges to simultaneously achieve increased market share and improved gross margins - a very difficult task. Sales people seek factors that win bids over competitive offerings - usually some trade-off between flexibility to tailor product to specific applications, better product appeal, higher performance and quality, and lower price. Manufacturing people are concerned about how to deliver high quality at low cost - usually some trade-off between standardized product configurations, range of customer applications and manufacturing processes. Service people are concerned about repair. Quality Function Deployment (QFD) and the associated "House of Quality" introduced by Donald Clausing in 1984 (Sullivan, 1986; Hauser and Clausing,1988), provide a systematic process for integrating and prioritizing desired "customer attributes" in the early design stages to achieve product descriptions with appropriate "design engineering characteristics".

However, customer requirements are not static, and refinements inevitably occur well after the product's concept phase. As the development team grows and progresses downstream, many of the added team members will not have had the opportunity for direct customer interactions. For example, manufacturing representation is normally limited (but should never be absent) at the outset of a new development program. Further, market environments often change because of competitive actions and other factors that modify the customer requirements ascertained at product initiation.

To cope with these changes requires the extension of the notion of QFD to Continuous Quality Function Deployment (CQFD). In Figure 6, CQFD is incorporated in the "Bottle" model as a "House of Quality" on wheels. This "House" traverses the "Bottle" model from Concept, through the detailed design phases of Product Design, Manufacturing Process Design, Production, and into the field. The attributes of the "House" gradually become more refined to satisfy the customer requirements that exist at the time of product introduction, rather than those that were envisioned at the time of program initiation. This notion may appear contrary to the demand for complete and stable customer product requirements prior to initiation of detailed design. It is not! Complete and stable requirements are mandatory before commencing the Product Design phase. Conversely, refinements are inescapable and must be managed as product development proceeds, even when development schedules are very short. Another challenge relates to deterioration in design intent

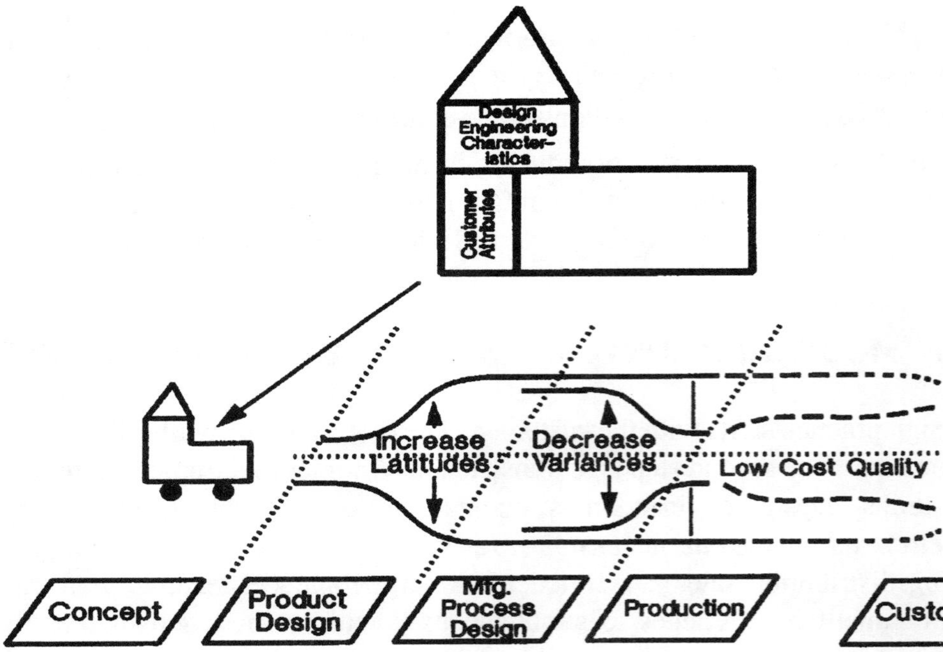

Figure 6: Control of Latitudes and Variances as the Product progresses through Product Design, Manufacturing Process Design and Production

latitudes and variances illustrated in Figure 6 as a product progresses through manufacturing and eventually endures the use and abuse of the customer. In manufacturing, variances can increase because of loss of process control, tool wear, and tight tolerancing.

Design engineers must interact with real customers in order to understand the real customer requirements for appeal, use and abuse of a product. Taguchi continually emphasizes that specifications entered into the House of Quality must be derived from customer use perspectives (field experience) rather than design engineering technical perspectives. In a recent paper entitled "Producer-Consumer Tolerancing", (Fathi, 1990) describes the difference between producer specification limits and consumer specification limits in terms of a model that is consistent with the Figure 6. The model extends the standard Taguchi assumptions to provide an improved representation of the relation of the increased variances experienced by the customer compared to those specified in manufacturing. Fathi's methodology can also be used to relate the larger variances that actually occur in manufacturing environments compared to the tighter specifications typically delivered by an engineering design team. To achieve customer satisfaction, specifications must be derived from customer use and abuse requirements, not from product development team perspectives about producer tolerances (manufacturing specifications). As Hauser and Clausing (1988) suggest, frequent face to face interaction with customers is the only way that a product development team can achieve sufficiently detailed information about customer requirements to bridge between engineering perceptions and customer experiences.

Unfortunately, the entrenched market research folklore discourages meaningful direct interactions between the customers and the people responsible for delivering the product. Traditional market research is performed by the "professionals", interpreted by the "professionals", documented by the "professionals", and communicated to other members of the product development team by the market research "professionals" in a fashion reminiscent of the old days when engineering threw the drawing over the wall to manufacturing. Coupling the Bottle and House of Quality aids the determination and management of product requirements that respond to all development phases, production and customer use.

CONTROLLING THE "BOTTLE" PARAMETERS: DESIGN PRACTICES

The "Bottle" defines a framework for integrating, managing, and communicating the full range of design practices appropriate during the detailed design process. These include Optimization, Design for Assembly, Design for Producibility, Functional Analysis, and Competitive Design Comparisons. Models like the "Bottle" will become increasingly important as the complexity of integrated sets of design practices continues to increase over time.

During the era of Frederick Taylor and Henry Ford, mechanical design engineering was already a sophisticated discipline. Design principles such as Loose Fit, Long Wear, Light Weight, Little Size, and Low Power were complemented with fundamental knowledge of thermal dynamics, mechanisms and structures. These apparently simple adages (they are not simple to implement) remain as powerful principles that sometimes get overlooked in the complexity of executing detailed design. For example, office appliances and many other products that weigh less usually cost less; specifically, cost reductions of copiers have paralleled weight reductions.

By the late 1960's, Siddall (1972) and others had established design processes on the firm basis of scientific method and analytical decision making. As early as the late 1970s, Boothroyd, et al (1982) stimulated significant progress in Design for Automated Assembly (DFA), which proved to be a major "process" step toward knocking down barriers between manufacturing and engineering. In the early 1980's, Ford, AT&T, and Xerox separately retained Genichi Taguchi in attempts to understand his design methodologies, which at that time had been utilized for two decades in Japan. Xerox has adopted Taguchi methodologies (Baker and Clausing, 1984; Taguchi; 1987; Byrne and Taguchi, 1987) as a cornerstone for its Design Engineering processes and Pugh Processes for Concept Selection (Pugh, 1981). Analysis, optimization (Reklaitis, Ravindran and Ragsdell, 1983) and other traditional engineering processes (Dieter, 1983) as the basic fundamentals of good Design Engineering. Figure 7 entitled, "Technical Processes Summary" is an overlay on the "Bottle" Figure 3; the entries in Figure 7 specify some exemplary design practices for Product Design and Manufacturing Process Design. The Concept and Production phases are not elaborated in the figure.

A less well known but very useful design practice is *Competitive Design Comparison.* It replaces emotion and opinion with a methodology for knowledge based evaluation processes of design approach alternatives. The premise is simple and powerful; it requires the design team to find the best competitive design that compares to their design task and either improve upon or copy that design. In order to demonstrate the merits of undertaking a new design, comparative evaluations with previous designs must be performed. The process provides strong motivation to create better designs than the world's best existing design. It also forces objective decision making about when not to do a new design.

CHANGE AGENTS FOR CONTINUOUS IMPROVEMENTS

Competitive Benchmarking shows that the best Japanese corporations invest a substantially larger portion of RD&E in Process Improvements and a smaller portion of Product Development than comparable U.S. corporations (Abbeglin and Stalk, 1985; Schnberger, 1982).

Figure 7: A Summary of Technical Processes (An overlay on the "Bottle" Chart, Figure 3)

	Product Design Processes	Manufacturing Process Design Processes
Quality		
	Competitive Design Comparison	
	Pugh Concept Selection	
	Design Optimization	
	Critical Tolerance Analysis	
	Taguchi Parameter Design	Taguchi Tolerance Design
	Design for Assembly	Taguchi Process Parameter Design
	Design for Manufacturability	Early Supplier Commitment
	Early Supplier Commitment	On-line Quality Control
Cost		
	Functional Analysis	Early Supplier Commitment
	Value Analysis/Value Engineering	Augmented Value Analysis/Engineering
	Design for Service	Critical Parameter Audit
		Functional Analysis
Schedule		
	First Instance Design/Quality	Supplier Development
	Rapid Prototyping	Prototype and Production
	CAE/CAD/CAM/CAT	Lead time reduction
	Product Delivery System	Manufacturing Planning
	Configuration Control	
	Design Quality Plan	
	Problem Management	

Consequently, Japanese Product Development processes, particularly Detailed Design Processes, have become so effective over the years that they can proliferate product lines very quickly with Product Development resources that are about one-fourth of those of a comparable U.S. firm.

Within U.S. corporations, the high costs and long schedules required to introduce new products cause management to prioritize investments toward the next new face-off product rather than toward the longer term topics of improving downstream Detailed Design Process. U.S. corporations do generally invest heavily in research and technology. However, these investments seldom reach to the downstream design issues about how engineers practice engineering design and whether they possess competitive skills, practices, and tools.

During my twenty or so visits to major corporations, an often cited procedure for improving product Quality, Cost, and Schedule was to establish "stretch" targets of improvement and then ask the Product Development team to develop the enablers to achieve the targets. This **management by exhortation** approach normally falls short. Even if a Product Development team creates process improvements, it is difficult to propagate the improvements to other Product Development teams.

Benchmarking of U.S. and Japanese corporations clearly indicates that the most successful corporations invest in change agents such as Engineering Design Institutes chartered to find, internalize, establish experts, create training documentation, train-the-trainers and train the workforce in the world's best Engineering Design processes. Such **Centers of Excellence** also exist in other disciplines, such as Marketing. A Marketing Center would be expected to find new processes, such as Quality Function Deployment, and rapidly spread them into the enterprise infrastructure using the same types of processes as cited for the Engineering Design Center.

Effective implementation of "agents of change" for continuous improvement can be characterized in terms of three process steps.

Recognition of the problem (or overcoming denial) is the first process step. It must start with a senior management team that recognizes that it is a major part of the problem and is prepared to both listen and lead during the difficult recovery period. Competitive benchmarking is a valuable tool for identifying **what** the problems are and even suggesting **how** to fix the problems. Benchmarking the **what's** - namely, Product outcomes - is fairly common. Benchmarking the **how's** - namely Processes - is harder and less common. A strength of benchmarking is that it replaces emotions and opinions with facts and data that help people to overcome the denial problem. Benchmarking is itself an important change agent (Camp, 1989).

Introduction of engineering design processes requires a dedicated team of technical practitioners working with enlightened leadership to penetrate the sturdy cultural barriers of 'status quo'; creation of the previously described Design Institutes and other Centers of Excellence requires modest investments which should be justified by the outcomes of the first process step of recognition.

This step is dangerous. Quality Engineering Design references how engineers practice design. Do they have skills in the world's best design practices? Do they utilize those skills when doing design? How are they managed? Does their supervisor measure progress by counting drawings completed, or does the supervisor measure progress in terms of improvement rates in Design Quality and the level of Design Quality verified prior to release of a drawing? Issues that reference this level of detail about how engineers practice Design Engineering are not commonly the forte of corporate senior management. Consequently, the engineering design subject often gets generalized to the broader topic of Product Development Processes where management is more comfortable; the issues of engineering design skills fade into the myriad of Product Development problems. When this occurs, the central point of this article becomes lost. Most aggressive corporations have developed adequate processes for managing product development. However, these Product Development Processes seldom correct the 'root-cause' problem of achieving world-class engineering design skills and practices.

A broken infrastructure is very difficult to mend. A friend noted that we are very good at shaking the trees and moving the monkeys around; however, such rearrangements do not solve the problems. We must change the ways the monkeys behave and the skills that they possess. The premier Engineering Design skills largely reside outside of the U.S. The U.S. must either bootstrap itself to competitive superiority or hire the world's best Design Engineers to lead the charge. Either approach requires investment in institutionalization of change agents for continuous improvement. Investments in process improvements are not in vogue. Most U.S. corporations are so starved for products that they continue to over-invest in new product offerings utilizing outmoded product development processes. An alternative is to invest in improving quality, cost and schedule Product Delivery Processes that enable product proliferation without increased resources.

Institutionalizing contemporary engineering design practices is a most difficult change. Most corporate leaders know what needs to be done. However, there are many, impediments. To maintain pace with global competition, enormous changes must be continually driven through enterprises. People discuss the need for improvements and at the same time resist the necessary changes. Even enterprises that have had enormous successes with contemporary design practices such as Boothroyd/Dewhurst in one portion of their organization continue to have difficulties in **institutionalizing** the new practices throughout their corporations. In Japan, successes are instantaneously integrated into the corporate culture. The single observation is a major obstacle for U.S. Corporations.

Another institutionalization problem relates to uniformity of processes across divisions. The common practice of holding each 'division' accountable for establishing its own world-class processes and producing world-class products leads to non-uniform practices throughout an enterprise and does not achieve world-class results. CEO leadership is required to create enterprise uniformity. Once this principle of "we are one company" is established, then the primary implementation process becomes **training, and more training** by people with world-class knowledge of their subjects.

Repair of the design engineering gap would enable a competitive resurgence of American corporations through their business and technical strengths. Business strengths include large local markets, strong sales, and service infrastructures, and low energy and raw materials costs. Technical strengths include world-class universities, software skills, extensive utilization of engineering automation and analytical tools for CAD/CAM, artificial intelligence, simulation and modeling, and individual creativity complemented by continuing in teamwork abilities. Corporate America has already proven its strengths just by surviving the broad consumer perception of inferior product quality compared with its global competitors. These strengths could propel a great competitive resurgence if corporate America could get its "House of Quality" in order by repairing the broken design engineering infrastructure.

Acknowledgments

This article is based on efforts of many people, too many to individually name on the Xerox team devoted toward improvement of engineering design capabilities. Extensive dialogue with Maurice Holmes, Xerox significantly contributed toward the formulation of the concepts put forth in this article. Jack Dixon and Mike Duffey, Associate of the Mechanical Automation Laboratory of the University of Massachusetts, made major editorial contributions to this article. Ken Ragsdell, of the University of Missouri - Rolla, has made numerous contributions over the years. Finally, the whole process of improving design capabilities with Xerox was initiated by Don P. Clausing in the late seventies, intensified in the early eighties with the hiring of G. Taguchi by Clausing and the creation of the Xerox Design Institute by Maurice Holmes and the author.

REFERENCES

Abegglin, J. C. and Stalk, G. (1985), Kaisha, The Japanese Corporation, Basic Book inc., 1985.

Barker, T. B., and Clausing, D. P. (1984), "Quality Engineering by Design: The Taguchi Method, " Proceedings of the 40th Annual RSQL Conference, 1984, pp. 1-23.

Bebb, H. Barry (1987), National Academy of Engineering, Edited by Phelps, P. B., "Responding to the Competitive Challenge, the Technological Dimension," The Bridge, Vol. 18, No. 1, Spring 1988, pp. 4-6.

Boothroyd, G., Poli, L. and Murch, L. E. (1982), Automatic Assembly, Marcel Decker, New York, 1982.

Byrne, D. R., and Taguchi, S. (1987), "The Taguchi Approach to Parameter Design," 40th Annual Quality Congress Transactions, 1987.

Camp, R. C. (1989), "Benchmarking: The Search for the Best Practices that Lead to Superior Performance," a series in Quality Progress, 1989.

Clark, W. D. (1990), "Bugs are Good: A Problem Oriented Approach to the Management of Design Engineering," Research Technology Management, May-June 1990, pp. 23-27.

Clausing, D. P. (1980), Personal Discussions; Currently, Bernard M. Gordon Adjunct Professor of Engineering Innovation and Practice, Massachusetts Institute of Technology.

Dieter, G. E. (1983), Engineering Design, A Materials and Processing Approach, McGraw-Hill, 1983.

Dixon, J. R. and Duffey, M. R. (1990), "The Neglect of Engineering Design," California Management Review, Winter 1990, pp. 9-23.

Fathi, Y. (1990), "Producer-Consumer Tolerances," Journal of Quality Technology, April 1990, pp. 138-145.

Harry, M. J. (1989), "The Nature of Six Sigma Quality." Government Electronics Group, Motorola, Inc., 1989.

Hauser, J. R., and Clausing, D. P. (1988), "The House of Quality," Harvard Business Review, Vol. 66, No. 33, May-June, 1988, pp. 63-73.

Holmes, M. F.(1985), Personal Discussions; Currently, Vice President and Chief Engineer, Development and Manufacturing, Xerox Corporation.

Ishihara, Shintaro, The Japan That Can Say No, Simon Shuster, 1989.

Jansson, D. G., and Smith, S. M. (1989), "Design Fixation," Preprints, 1989 NSF Engineering Design Research Conference, College of Engineering, University of Massachusetts, Amherst, June 11-14, 1989, pp. 53-79.

Pugh, S. (1981), "Concept Selection - a method that works," Proceedings of the International Conference on Engineering Design, Rome, March 1981.

Reklaitis, G. V., Ravindran, A., and Ragsdell, K. M. (1983), Engineering Optimization Methods and Applications, John Wiley and Sons, 1983.

Schonberger, R. J. (1982), Japanese Manufacturing Techniques, The Free Press, 1982.

Siddall, J. N. (1972), Analytical Decision Making in Engineering Design, Prentice Hall, Englewood Cliffs, N.J., 1972.

Sullivan, L. P. (1986), "Quality Function Deployment," Quality Progress, June 1986, p. 39.

Taguchi, G. (1987), System Experimental Design, Engineering Methods to Optimize Quality and Minimize Costs, UNIPUB/Kraus International Publications, 1987.

THE UNIFIED METHODOLOGY ON DESIGN FOR LATITUDE

AS A CONCEPT AND CAPABILITY FOR TOLERANCE ANALYSIS AND SYNTHESIS

Jean M. Parks*

SUMMARY

Despite widespread development of computer-aided-engineering tools, tolerance analysis and synthesis capabilities are still in their infancy. Additionally, at present, effects of variability are mostly addressed through tedious Monte Carlo simulation. These shortfalls significantly impact not only efforts to improve design quality and engineering productivity (which affect product quality, cost, and delivery cycle), but also the design process itself.

To overcome these deficiencies, a Unified Methodology on Design for Latitude (UMDFL) was developed (Parks, 1991). Based on a stochastic modeling approach coupled with a mathematical technique, the UMDFL provides a rigorous quantitative engineering capability that captures the effect of variability without use of Monte Carlo. The UMDFL unifies the flow of variances from the most elementary piece part level to the system performance level, and, further, includes Taguchi Quality Loss considerations. It therefore stochastically links customer satisfaction/dissatisfaction to product design. The UMDFL enables (1) a comprehensive tolerance analysis capability well beyond that for typical geometric tolerance analysis, and (2) a tolerance synthesis methodology that is based on its enhanced optimization techniques. Further, the UMDFL puts the technology and product delivery processes in a proper perspective with all other processes and tools needed for the realization of design intent.

In this paper, the UMDFL has been recast as a unified concept and capability for tolerance analysis and synthesis. The methodology emphasizes the need for the fundamentals of engineering mechanics, infrastructure enablers, and the requisite design engineering work processes, including robust design practices, to support its full implementation.

* Research Fellow, Xerox Corporation, Webster, NY 14580

DEFINITIONS

Latitude: Latitude refers to the amount of variability that can be tolerated with negligible impact on the product's performance. The larger the latitude, the more robust the product. Therefore, throughout technology and product development, latitude is to be expanded to reach a target level through selection and control of critical parameters.
Critical Parameters (CP): Those parameters which are determined to be critical to the product's function. Typically, they are at a higher level than that of piece part dimension, and are controllable during technology and product development (e.g., forces, temperature, relative positions, materials, etc.). Often, some of these are the "control factors" in Taguchi methods.
Initial Variability: Variability during manufacture (as distinguished from that from product degradation due to usage).
Total Cost to Customer (TCC): Customer's cost to own (or lease) a product and operate, maintain, and have it serviced.
Total Loss to Customer (TLC): Service cost and cost of loss of use when a product (machine) fails to operate.
FOV = Flow of Variances
SMB = Stochastic Modeling Based
UMC = Unit Manufacturing Cost
DFX = Design Practice for achieving better X (Assembly, Manufacture, Maintenance, Service/Repair, Disposal/Recycle, etc.)

INTRODUCTION

At Xerox, an in-depth study on how to significantly improve productivity of design engineering was conducted last year. One of the glaring deficiencies found was that current methods of tolerance analysis and synthesis, and the practice thereof during product development, were inadequate. This was not a problem unique to Xerox because similar deficiencies had been exposed in a recent ASME/NSF report (Tipnis, et. al., 1989). It is ironic that in this age of sophisticated technologies, tolerance analysis is in its infancy. How, indeed, are industries able to implement such technologies into products? The answer, of course, is by reworking designs repeatedly at the expense of lower product quality, higher cost, larger resources, and longer time-to-market. With inadequate tolerance analysis and synthesis capabilities, the effort required to deliver a complex product such as a copier is indeed immense.

The tolerance analysis situation stated above refer to geometric tolerances only. The larger issue of linking variability from sources beyond geometric tolerances and addressing the effect of all variabilities on performance has received very little attention. Although computer aided engineering (CAE) tools have become commonplace, with the exception of a few commercial geometric tolerance analysis packages, these tools enable only deterministic quantitative engineering. Simulation by super-imposing the Monte Carlo technique onto deterministic models was still the way effects of variability were investigated. Often this is inadequately or inefficiently carried out to yield required accuracy.

As for tolerance synthesis, it is in fact a problem of allocating piece part variabilities and consequently, a design optimization problem. Depending on the level of optimization, from local to global, the tolerance synthesis activity yields corresponding levels of power in cost to performance trade-offs. Put into this context, the problem of mechanical tolerancing, which consists of both tolerance analysis and tolerance synthesis activities, becomes the total activity of engineering a product. The latter has been referred to as the "product development (or delivery) process" in recent years. However, prescribing all activities that are needed for product delivery loses the focus on the engineering aspects at the working level.

The Unified Methodology on Design for Latitude (UMDFL) (Parks, 1991) was developed to enable understanding how every variability affects performance and how the pertinent variabilities are to be controlled within reasonable cost and performance targets (i.e., optimized); it is a "womb-to-tomb" approach to design quality and optimization through addressing variabilities. Presentations of the UMDFL have been well received in the design community (e.g., the Design Productivity International Conference held in Honolulu in February 1991). Because the UMDFL covers many of the open issues identified by the ASME/NSF International Workshop on this subject (Tipnis et. al., 1989), we have lifted its relevant elements and recast them as a concept and capability on mechanical tolerancing. Indeed, in this paper almost all of the major issues of mechanical tolerancing are addressed.

The Essential Elements of UMDFL:

The UMDFL starts with a rigorous definition of the levels of variabilities and what is needed to deal with them. The resulting chain clarifies the heretofore missing links that lead to quantifying performance variability. Next, performance variability is linked to customer dissatisfiers. Understanding the nature of these dissatisfiers is essential to relate manufacturing variability and degradation (and therefore, the inability to meet planned design latitude) to Quality Loss. Since variability and degradation can be dealt with by stochastic models, then linking to service data functionally connects the entire chain, thereby providing a coherent approach to design analysis and optimization of the entire system. A mathematical technique (Parks, 1980-1982) for obtaining cumulative distributions obviates need for using Monte Carlo and forms the foundation of the UMDFL technique.

The work content and infrastructure enablers for the UMDFL process are described to enable understanding levels of tolerance analysis and synthesis that can be conducted. Further, relationship to technology and product delivery processes, in particular, to the design and engineering processes, is explained. The UMDFL includes the following:

- Flow of Variances (FOV)
- Relationship to Technology and Product Delivery Processes
    - Modeling and Experimentation
    - Defining Critical Parameters; Optimization
    - Design Analysis (includes Tolerance Analysis)
    - Manufacturability Analysis
    - Driving the Required Work Processes
- Infrastructure Requirements
- The Core Tools and Techniques; Application to FOV
- Putting in Perspective Work Processes and Tools

VARIABILITY: FROM LOWEST LEVELS TO PERFORMANCE LEVEL

For competitively marketed complex electro-mechanical products such as copiers, a major challenge of technology development and design implementation to achieve specified function is how best to deal with variabilities due to manufacturing, customer noise, and usage degradation. Products such as aerospace and nuclear systems can afford redundancies and standby systems to handle degradation and very tight tolerance delivered by high cost precision processes. Such are generally not cost-wise practicable to the medium-weight product segment. Further, products such as copiers contain complex and highly interactive subsystems which are sensitive to their interaction with the materials being processed (such as toner and paper) as well as to variabilities in the user environment. Thus, how to quantify and minimize the effects of manufacturing variability, customer noise, and degradation is essential to understanding the quality of product design. This ability to quantify the effects depends on understanding the proper chain of relationships, without missing links, that articulate the flow of variances. Figure 1 shows the proper chain (though not a linear chain, the term "chain" is used here because we associate its elements with "links"). Shown in italics, next to the boxed entities, are the key software tools or processes for quantifying the various aspects of variability. The Tolerance Analysis and Failure Rate Prediction tools are the core UMDFL tools provided through our Stochastic Modeling Based (SMB) approach.

Though not user friendly and only sparsely used, commercial geometric tolerance analysis tools exist, but with strong limitations. And databases are still lacking adequate content and accessibility for tolerancing needs. Note that in Figure 1, only the 3 boxes on the top left hand side and their links relate to geometric tolerancing that currently is being addressed by CAD/CAE and manufacturing industries. Note also that although material property variability is a part variability, it is shown separately because its geometric variability is less significant in the chain. What has heretofore not been clarified are the Assembly to Critical Parameter link, the links to and from Material Property Degradation and Component Degradation, and the links from the 5 entities to Performance. Engineers are generally aware of all contributors to performance variability, but the FOV had not been articulated in this specific chain to facilitate their determining the impact quantitatively. For example, except for a trivially simple system, how would the engineer determine that his assembly could deliver an appropriate force distribution under dynamic conditions so that performance within the specified acceptable limits could be attained. The UMDFL tools and techniques introduced here provide such capability.

Figure 1 represents the flow of variances at either subsystem or system levels of any hierarchically complex system. The "component" at the lowest level could be an electrical or mechanical component (sometimes referred to as a part). At some next level, "component" could be a subsystem to a larger subsystem or system whose performance variance is to be analyzed. Sometimes, a component may affect critical parameter values of a subsystem, in which case, in Figure 1, arrows could link the "Component V" to the "Critical Parameter V" as well.

Further, in Figure 1: (1) The hierarchical structure of entities need not be strict. For example, a piece part dimension may itself be a Critical Parameter. Similarly, a material property may well directly appear as a random variable in the stochastic model for calculating the expected performance. The important point is that there is a proper hierarchical flow for stepwise quantification of the pertinent effects of variances, and (2) While performance is represented for the total system, in practice, quantification is typically carried out at the subsystem level; performance at the total system level is done by summation/aggregation. However, where, say, 2 subsystems interact, appropriate prior local analysis or optimization for each would fix some Critical Parameters (CPs) and their variances, leaving a selected set of CPs to be analyzed/optimized for both subsystems together via the UMDFL process.

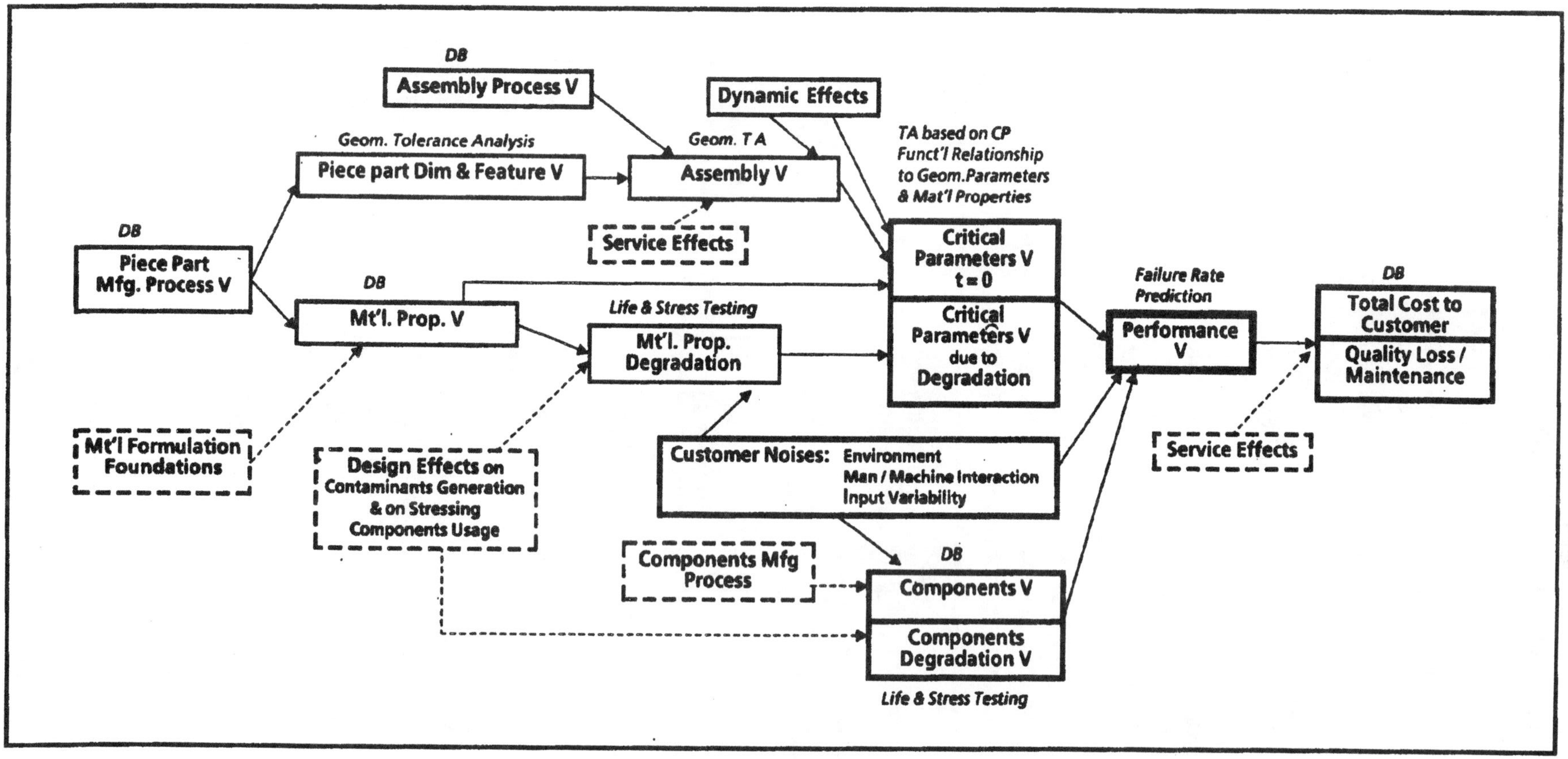

FIGURE 1. Flow of Variances (V)

(dotted box) Additional factors which affect variances

DB = Databases; CP = Critical Parameters; TA = Tolerance Analysis

TOLERANCE ANALYSIS AND TOLERANCE SYNTHESIS

Tolerance analysis and tolerance synthesis comprise the subject of tolerancing. As shown in Figure 2, the directions of evaluation of variances differ for analysis and synthesis according to the goals, performance analysis and optimization, respectively.

Tolerance analysis is concerned with analyzing the effect of individual variances and deviations on variability of the Critical Parameters; e.g., the resultant variability of a force, for which an assembly was designed to deliver, can be analyzed. Performance analysis carries on beyond that level to determine the effect of all variability contributors on the system performance. The tolerance analysis and performance analysis mode is in the direction of the arrows shown in Figure 1.

Tolerance synthesis involves allocation or distribution of variances to individual components based on system requirements. For example, if a certain permissible range for the force is to be controlled within a specified permissible variability for the assembly, tolerance synthesis methods would specify the permissible piece part variability while taking into consideration the degradation and the cost factors. At this level, local optimization is being conducted, i.e., trade-offs can be made amongst the part dimensional tolerances and material property variability under the constraint of the permissible Critical Parameter range for achieving required latitude.

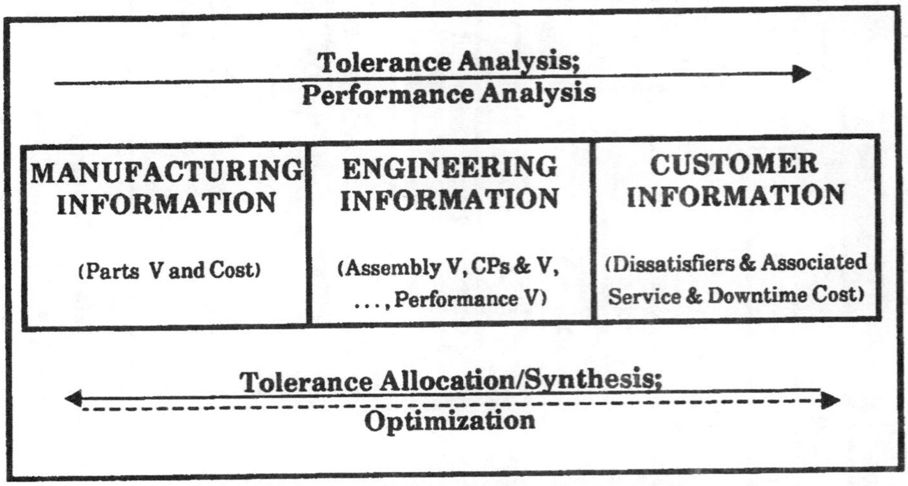

Figure 2 Analysis and Optimization Processes

Additionally, the UMDFL process and technique allow tolerance synthesis to be performed at subsystem and system levels of optimization. For such optimization, the performance level is treated as a constraint. Allocation is performed by following the chain to the left in Figure 1 so as to optimize the manufacturing tolerances/cost through trade-off analysis. It is also possible to optimize tolerances against total cost to customer and Quality Loss as shown later.

Even with the UMDFL tools, optimization currently requires a "what-if" analysis process (hence, the dotted arrow is shown to support the tolerance synthesis direction in Figure 2). Although enhanced synthesis capability may be developed, it will not replace the important role of tolerance analysis, applied to the level of Critical Parameters, for verification of design quality.

Nature of Customer Dissatisfiers and Service Calls:

In Figure 3, the typical product related customer dissatisfiers of a copier are shown. To truly optimize requires understanding the causes for service calls, their relationship to parts variability or degradation, and the associated service cost and dollar loss due to loss of use (e.g., machine downtime at a quick print shop). For the most part, these causes can be linked to specific dissatisfiers which fall into the categories of 'soft' and 'hard' failures as shown in Figure 3.

Note that most of the dissatisfiers in Figure 3 are due to 'soft failures'; these are mostly due to (a) Critical Parameter and component degradation (drifting their values/responses beyond the

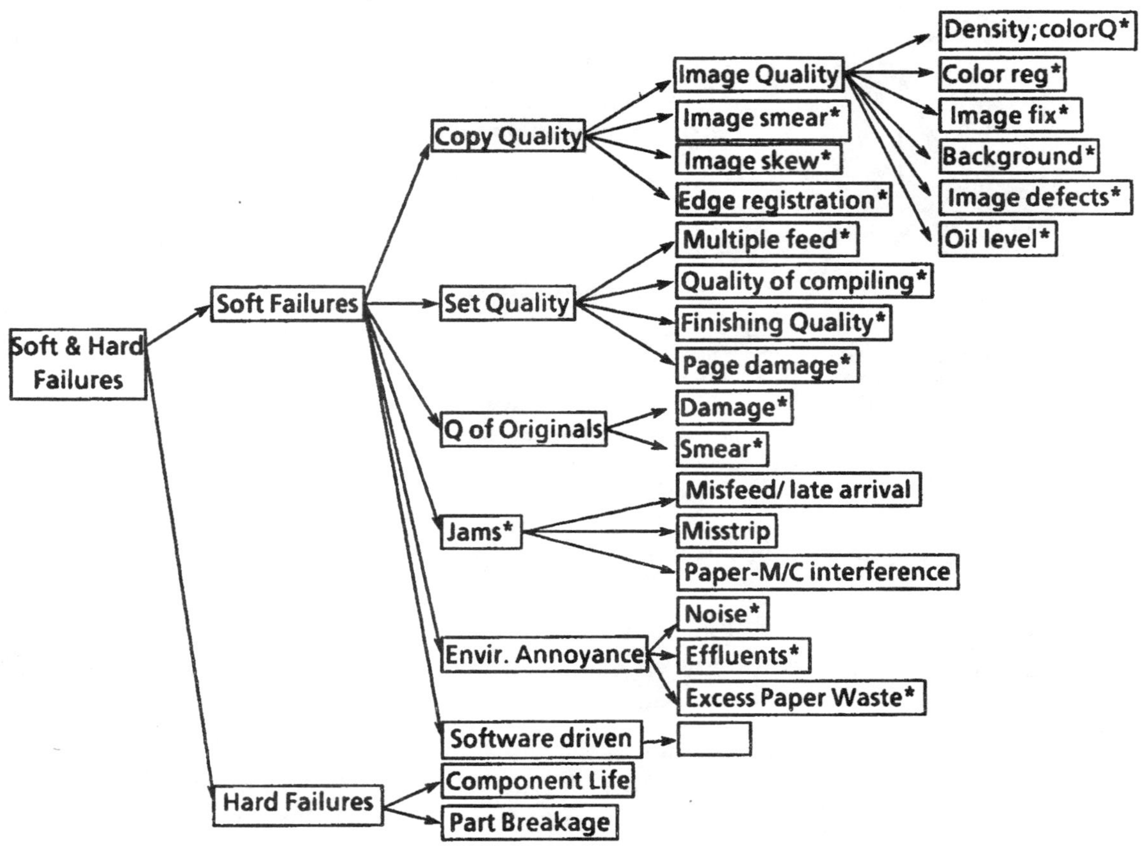

Figure 3 Classes and Elements of Customer Dissatisfiers
(Elements with asterisks are at levels that a customer typically sees.)

range of design intent, i.e., beyond the range that would deliver the planned performance latitude), and (b) customer noises beyond planned stress levels. Soft failures, however, are characterizable by stochastic modeling which happens to work well with the concept of Quality Loss. For each dissatisfier, one would determine the point at which a customer would call for service to correct a soft failure and the more appropriate set of tolerances such that Quality Loss would be reduced according to the product quality goal.

Each hard failure results in a service call. However, these generally represent a small percentage of the total service calls. Since the Quality Loss Function can at best be grossly applied here and preventive maintenance often takes care of many impending component hard failures, the Quality Loss discussions in this paper will be limited to soft failures. Note that symptoms of some hard failures may appear as soft failures (e.g., customer will see a coronode breakage as a copy quality defect, but such defects typically would appear as a step function degradation).

Use of Quality Loss for Product Design Decisions:

The concept of Dr. Taguchi's Quality Loss Function (Phadke, 1989) states that the penalty cost (Quality Loss) of deviating from the target within the limits of customer tolerance increases quadratically with the deviation from the target. The Quality Loss Function, therefore, is to be used for determining an appropriate trade-off point for any characteristic (performance, part, etc.) prior to reaching the customer tolerance limit, $M+\Delta C$, shown in Figure 4. The quantifiable Quality Loss

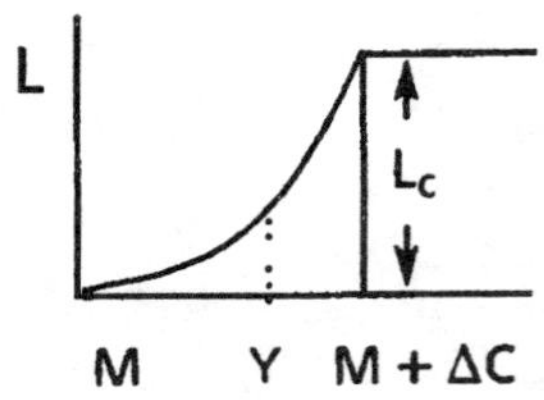

Quality Loss Function $L = L(Y)$
ΔC = customer tolerance
L_c = cost of exceeding customer tolerance

$L_c = K(\Delta C)^2$ → $K = L_c / (\Delta C)^2$

K is known (associated with point of needing a service call)

Q Loss at Value Y

$L = K(Y-M)^2$

Example: Suppose cost of service + cost of loss of use is L_c = \$160 for customer tolerance of $\Delta C = 4$ (units from mean M).
Then $K = 160/4^2 = 10$.
The loss at $Y = M+2$ is then: $L = 10(2)^2$ = \$40

Figure 4 Quality Loss Function and Total Cost of Ownership

is generally associated with a specific failure mode of operation (e.g., for a copier, increase in the rate of misfeed or late-feed, causing 'soft' jams) for this can be associated with a service call (see Acknowledgment).

Per UMDFL method, for any customer dissatisfaction factor, e.g., late-feed, all critical parameters and noises which cause this feeder problem are identified and traced down to the variabilities of all parts involved so as to determine the trade-offs between the tolerances to optimize their selection. In other words, optimization is carried out within the context of the coherent whole set relative to the specific dissatisfier of that subsystem. UMDFL derives its strength from this quantification capability through which true implementation of Taguchi Quality Loss Function can be achieved to optimize design decisions. The steps involved in this process are as follows:

• Determination of ΔC is key: It is associated with a dissatisfier and therefore relates down the chain to an entire set of ΔCs (parts tolerances) associated with the subsystem generating this dissatisfier. Figure 5 shows the chain from Customer Dissatisfier to the associated part tolerances (note that ΔC is specific to a planned product; e.g., a customer is more likely to place a service call if a multifeed (feeding more than one sheet during an actuation) rate is two per thousand feeds in a high volume copier than in a low volume product).

• Marketing must know or estimate the loss L_C associated with this dissatisfier from the field.

• The Loss L for a dissatisfier is the appropriate level of Quality Loss commensurate with the product quality to be offered. The latter is dependent on how the total optimization problem considered performance and cost targets, the expected usage, and the associated service strategies for the product. L is to be used as a constraint for the design and for establishing the maintenance process.

• Using the Loss L as constraint, trade-off decisions on the set of Critical Parameter tolerances and component quality can be made for the subsystem and thence, on the part tolerances. By selecting the Loss constraint at a yet smaller value provides safety factors or opportunities to satisfy customers beyond their expectation for the product.

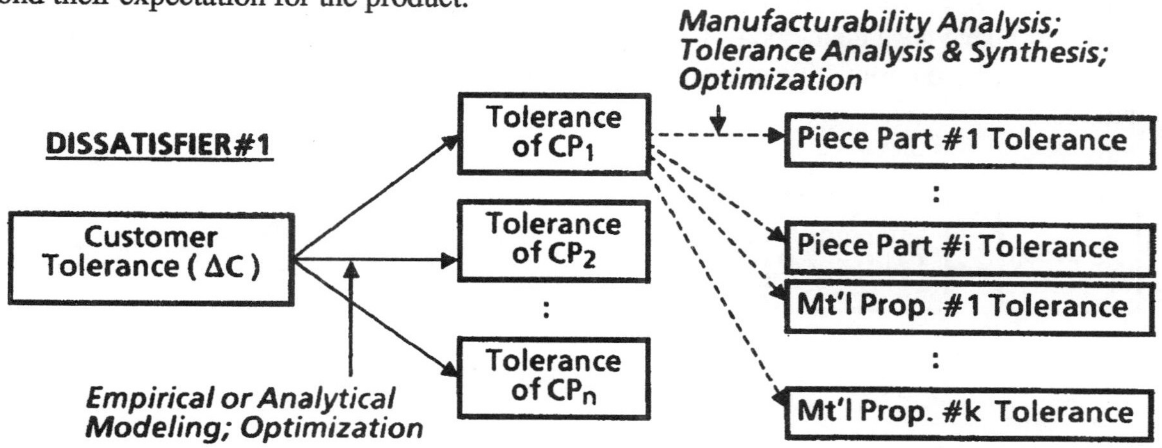

Figure 5 Customer Dissatisfier to Lowest Level of Tolerances

The Double Ended Constraint Optimization Problem:

Although difficult, true optimization is possible because of the total analysis capability of the UMDFL method. However, the over-constraint typically placed on the product design synthesis team violates any optimization capability. To illustrate this, refer to Figure 6, which is a simplified schematic of Figure 1 combined with Figure 2, where Total Cost to Customer (TCC) has identified Unit Manufacturing Cost (UMC) as a significant constraint. This UMC may contradict the optimized set of manufacturing tolerances that were optimized against a Quality Loss constraint. To highlight the importance of considering both factors for optimization, the TCC and the Total Loss to Customer (or Quality Loss) are shown as components of the same box at the right hand end of the chain in Figure 1. The typical optimization problem involves either delivering the minimum Quality Loss at the planned UMC or delivering a minimized UMC at an expected Quality Loss level. When both Quality Loss and UMC are constraints, the optimization problem is over-constrained; only a minuscule chance exists that the optimized set of tolerances would satisfy both constraints (except when the UMC constraint value chosen was a good approximate to begin with). Presently, only a combination of clever engineering and additional time/resources allow meeting performance level within a reasonably modified UMC. Clearly, only after conducting proper product tolerance synthesis and quantifying the effect of tolerances on Quality Loss could there be final optimization and commitment to specific product quality and UMC. Perhaps for each product, the relative priority of UMC and quality goals should be pre-established so that one of these can be considered as a non-stringent constraint to optimize the other within a reasonable range (in essence, this is the principle of Taguchi Methods).

THE STOCHASTIC MODELING BASED TECHNIQUES AND TOOLS

The commonly used techniques for tolerance and variability analysis are: (A) Root-Sum-

Square (RSS) and (B) the Error Analysis Technique (also referred to as Taylor Series, Linearization, or Error Propagation Technique). Also, Monte Carlo technique is commonly employed to simulate the impact of variability. As a part of the task of developing mathematical techniques and tools for evaluation of every link of the FOV in the UMDFL, the Stochastic Modeling Based (SMB) techniques (C) have been developed. The fundamental mathematical technique for (C) provides a new way to calculate cumulative probability distributions, and its extensions enable the UMDFL capabilities.

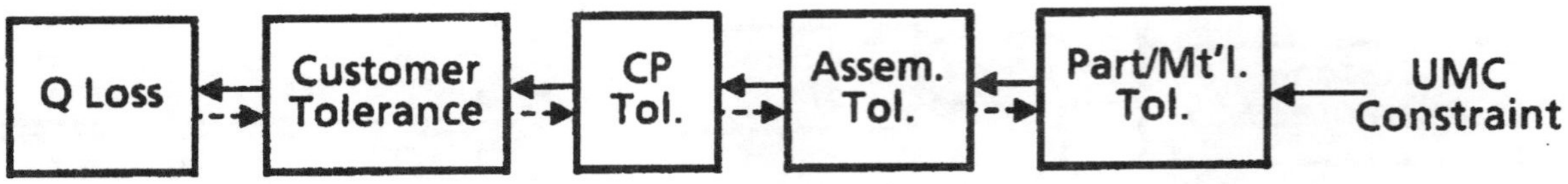

Figure 6 Optimization Problem -- Over-Constrained?

Background:

Over a decade ago, Xerox realized the importance of exploring the relationship between input variabilities and output performance expectations of copiers. In 1980, the author began developing a failure rate prediction methodology based on probabilistic modeling of first principles. Due to the author's occupational change in 1982, this work remained idle until 1990 when it was restarted partly because of realizing that this approach could be extended to apply not only to mechanical tolerance analysis, but also, to the entire subject matter of variability analysis and optimization for the UMDFL. The approach, as described below, starts with the method for failure rate prediction. It is then shown that it is essentially a process for calculating the cumulative distribution of any resulting "situation" that is dependent on input distributions.

Stochastic Modeling:

Suppose the drive force, F_D, for a paper feeder is represented by the following deterministic model: $F_D = (\mu_{EP} - \mu_{PP})F_N$

where F_N is the normal force, μ_{EP} the elastomer to paper coefficient of friction, and μ_{PP} the paper to paper coefficient of friction. Assume that F_N is dependent on assembly variables X_1, X_2, & X_3 and on the dynamic effects of actuation given by a function $G(X_3, X_4)$. Then assuming that μ_{EP}, μ_{PP}, $X_1, \ldots, X_4$ have statistical variability, the stochastic model for F_D is the same expression, but F_D is now the resultant random variable dependent on these 6 random variables.

The Fundamental Mathematical Technique:

A mathematical technique for handling stochastic models was developed in 1980 when the author derived a comprehensive theory of failure rate prediction. To illustrate this method, consider a characteristic performance variable, ξ (e.g., the time of arrival at a certain point along the paper path). ξ is a function of N random variables. Then for some region Ω of the N-dimensional space, the failure condition

$$\xi(X_1, X_2, \ldots, X_N) > \xi_0$$

is satisfied. The mathematical approach was to use this failure condition to determine the failure region, Ω , and integrate over it the joint probability density function of the N random variables to compute the probability of being in that region, thus obtaining the probability of failure,

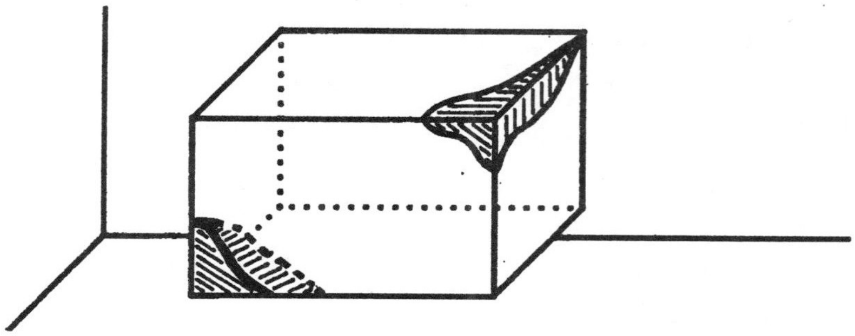

Figure 7 Failure Region Ω is Shaded
(Schematic depiction of N-dimensional space as a 3-dimensional cube)

$$P(\text{Fail}) = \int_{\Omega}^{\cdots}\int f(X_1,X_2,\ldots,X_N)dX_1dX_2\ldots dX_N$$

where f( ) is the joint probability density function of $X_1,X_2,\ldots,X_N$.
What this entails is an inversion of the failure condition, $\xi > \xi_0$, to yield,

$$\{X_1 : X_1 < H_{12}(X_2,\ldots,X_N)\}$$

or $$\{X_1 : X_1 < H_{12}\} \cup \{X_1 : X_1 > H_{21}(X_2,\ldots,X_N)\}, \text{ or......}$$

where H_{k1} and H_{k2} are surfaces bounding the kth subregion of Ω . In this way, we obtain:

$$P(\text{Fail}) = \sum_{k} \int_{X_N}\ldots\int_{X_2} \int_{H_{k1}}^{H_{k2}} f(X_1,X_2,\ldots,X_N)dX_1dX_2\ldots dX_N$$

Note that when there is one failure condition, it defines the limits of integration that differ from the full range for one of the random variables. Extensions to > 1 failure condition were carried out, yielding multiple integrals with limits of integration differing from the full range for ≥ 1 of the random variables. Appropriate manipulation of the failure conditions, represented by the *system* $\xi > \xi_0$, can facilitate a better representation of the surfaces bounding the failure regions than some other alternative ways of inverting the system of conditions, and may reduce computational time.

The same technique was also applied to address failures in the system sense, i.e., when it is a jth failure mode in a system of failure modes. The technique applies as well to cases with non-closed form failure conditions. Following the FOV chain in Figure 1, it is possible to see how deterioration and customer noise come into play in the P (Fail) equation stated above. Further, the way to define operating windows was extended, by not only taking into account probabilistically feasible regions for a single failure mode, but also the trade-offs of the failure rates of different failure modes of the same subsystem and their associated Critical Parameters and tolerances.

The above would not have been meaningful in practice except for a small number of random variables if a computational approximation technique were not also developed to significantly reduce computational time for multiple integration. This paper, however, will not cover the approximation technique we developed. Quadrature techniques exist in literature; some are adequately efficient for simpler UMDFL applications given today's workstation speed.

A Method for Calculating Cumulative Distributions:

The failure rate prediction method above extends very simply to a method for calculating cumulative distributions since P(Fail) = Prob($\xi < \xi_0$) is the cumulative distribution of ξ at $\xi = \xi_0$. Note that the multiple integration representation holds for any type of input and resulting distributions. Here, $\xi < \xi_0$ can be taken to represent a system of, say, m conditions and thus the cumulative distribution is the probability of a "situation" arising (such a joint distribution could be represented as an m+1 dimensional plot). Having this means of calculating the cumulative distribution of any random variable means it is but a simple extension to conduct analyses which are dependent on probability distributions, including tolerance analysis. Because this technique is many orders of magnitude faster than Monte Carlo, it eliminates the need for using the latter; this brought feasibility to implementing the UMDFL concept. The UMDFL core tools consist of a set such that a best one exists under any possible situation for each link of the Flow of Variances chain.

The Process:

The process of determining failure rates can be illustrated by considering a simple example of a failure mode in the image transfer region of a copier. Suppose that the failure conditions are given by

$$Z < Z_0, \text{ and } \theta > \theta_0$$

where θ_0 is a constant and Z, Z_0, and θ are variables derived from modeling (analytically or empirically) the system pertaining to performance in this region. Suppose that Z and θ were obtained as closed form expressions of N_1&N_2 variables respectively, and Z_0, which depends on N_3 variables, is obtained via a computational scheme. The set of $N_1 \cup N_2 \cup N_3$ variables is then $X_1, X_2, \ldots, X_N$ where some of the X_ns are dependent on the paper type (R_1), the usage environment (R_2, R_3), and use/age = "time" (R_4). Then using the failure conditions, one can obtain the failure region by finding the surfaces bounding it. Suppose in this case, there is only one contiguous failure region and the *easiest* inversion process yields limits of integration on X_1&X_2, which are surfaces, as indicated below, in the N-1 dimensional space and the N-2 dimensional space respectively, then we obtain the probability of failure for this set of usage conditions, (i.e., represented by the arguments $R_1, \ldots, R_4$),

$$P(R_1, \ldots, R_4) = \int_{-\infty}^{+\infty} \cdots \int_{-\infty}^{+\infty} \int_{\hat{X}_2}^{+\infty} \int_{-\infty}^{\tilde{X}_1} f(X_1, X_2, \ldots, X_N) dX_1 dX_2 \ldots dX_N$$

Integrating this over the distribution of market use and "time" gives the actual failure to this time; dividing by this time gives the average failure rate to this time. Note that the density distributions are dependent on the argument values $R_1, \ldots, R_4$ chosen, thus allowing us to determine their effect on performance accordingly.

Extension to Variability Analysis and Tolerance Analysis:

Variability analysis consists of determining the resulting distribution of the output parameter due to variability of the input parameters. As we stated, the mathematical technique shown yields a closed form expression for the cumulative distribution. From this, the density distribution is easily obtained also. If the limits of integration of the multiple integral representation are closed form expressions, then differentiation of the cumulative distribution will yield a closed form expression for the density function. If the limits of integration include accessing non-closed form terms, then numerical differentiation of the cumulative distribution will yield the density distribution. When the output variable is dependent on a large number of input random variables, segmenting these into groups to enable an m-step evaluation process is often possible. For this, calculating density distributions is necessary.

The method to conduct tolerance analysis is straight forward. Since we have the method to calculate the cumulative distribution at any ξ_0, it merely involves selecting the specific values of the cumulative distribution equivalent to, for example, those at the $\pm 3\sigma$ points from the mean for normal distributions (i.e., 0.0013 and 0.9987) and executing the calculation while coupling it to a root-finding scheme to obtain the two associated ξs. The difference between these ξ values and the mean would be what one typically identifies as tolerances (i.e., tolerances today are defined as the $\pm n\sigma$ limit values required for a design (n is typically 3 today)). Sometimes tolerance analysis is approached through analysis of the probability of success.

Extension to Operating Window Optimization:

To determine a "Failure Rate Dependent Operating Window" (Parks, 1980-82, 1991; Clausing, 1981-82), consider all major failure modes of a subsystem that have dependence on a common subset of important variables, say, $X_1 \& X_2$, affecting the performance of the subsystem. Keeping those as arguments, the failure probabilities are calculated for each selected failure mode. This allows selection of the best region for the best combination of failure rates for the set of failure modes through trade-off. Note that for any of the random variables, varying the density distributions yields different failure rate surfaces in the X_1X_2-space, thus providing an opportunity to trade-off both failure rates of the different failure modes and the tolerances of critical parameters, and therefore, the parts tolerances. Additionally, even though $X_1 \& X_2$ were kept as arguments, the effect of changing their distributions on the failure rate surfaces in the X_1X_2-space can be calculated by taking the mean values of $X_1 \& X_2$ as the arguments, while varying their distributions when evaluating the multiple integrals by changing their standard deviations. In this way, the effect of varying their distributions on the failure rate surfaces can also be brought into the trade-off evaluation. This process can be extended to operating windows with any number of variables.

COMPARISON OF TECHNIQUES FOR TOLERANCE AND VARIABILITY ANALYSIS

In Table 1, a comparisons of limitations and requirements of the RSS, the error analysis and the SMB techniques are shown. In Table 2, the tools and techniques applicable in each of the situations are shown. Note that the SMB technique is applicable to virtually all situations except when no functional relationships exist.

The only occasions where the error analysis technique may be preferred are when one wishes to quickly approximate manually a normal distribution for the resulting random variable, or where computational time with the SMB technique might be too long, such as when small computer systems are used to compute very large number of input random variables that cannot be segmented into subsets for analyzing in a nested process with the SMB technique in ≥ 2 steps. Table 2, then,

TABLE 1. TOLERANCE / VARIABILITY ANALYSIS TECHNIQUES

TECHNIQUE	APPLICATION CONSTRAINT
A. RSS	• Currently used by Designers for Geometric tolerancing • But as currently practiced, most who use it use it for worst case (i.e., not RSS, but "Stack-Up") • Is sensible only for linear assemblies • If linear, & if input RVs are normal, can use basic probabilistic properties to calculate the resulting RV's normal distribution (**) • If not linear or all normal, especially not appropriate for small # (e.g.,~ $\leq$ 5) of part dimensions
B. Error Analysis Technique (or referred to as Taylor Series, Linearization, or Error Propagation Technique)	• Basic technique for current commercial TA packages for geometric tolerance analysis (some, however, use Monte Carlo) • Requires functional relations (Commercial S/W builds on its parametric capability to automate it for geometric tolerancing) • Not appropriate for non-normal resulting RV of small # of part dimensions • For larger # of part dimensions, can use (**) to approximate normal distribution for the resulting RV; acurracy is less than if using (C). 6σ quality would require accuracy beyond using this technique (NOTE: In A, use of (**) is same as here except that there sensitivity coefficients are taken to be 1) • Can be used for nongeometric tolerancing, but with the same constraints listed
C. Stochastic Modeling Based (SMB) Technique	• Requires functional relations • No constraint on: -- Geometric or nongeometric -- Distribution types -- # of RVs (if # of RVs is too large and thus computatonal time would be too large, can use process of "m-step segmentation" of relationships to successively calculate distributions; if not possible, use approximation by normal distribution and use B.) • Extends to performance analysis capability and beyond

NOTE:

- Computationally, C is not via using Monte Carlo; it is far more efficient
- C far more readily enables doing tolerance specification and optimization due to its lack of restrictions; optimization is via the UMDFL umbrella which calls for the databases and supporting tools to complement C.

TABLE 2. APPLICABLE TECHNIQUES for FOV CHAIN

(Letters refer to techniques in TABLE 1)

INPUT RVs	RV RELATIONSHIP	RESULTING RV	PP → ASS'BLY	MT'L PROP → CP ASS'BLY → CP	MT'L PROP → FR COMPON. V → FR CP → FR
EACH RV NORMAL					
• ≤ 5 RVs [a]	• Additive	• Normal (N)	A ; B ; C	B ; C	C" ; @
	• Non-Additive*	• Not N	C	C	C"
	• Non-Additive***	• Not N	A (rough)	---	---
• > 5 RVs	• Additive	• N	A ; B ; C	B ; C	C" ; @
	• Non-Additive*	• Not N;	C	C	C"
		• If approx. with N	B ; b	B ; b	@ ; C"
	• Non-Additive***	• Not N; approx. with N	(1) (rough)	---	---
RVs NOT ALL NORMAL					
• ≤ 5 RVs	• *	• Not N	C	C	C
	• ***	• Not N	A (rough)	---	---
• > 5 RVs	• *	• Not N;	C	C	C
		• If approx. with N	B ; b	B ; b	@ ; C
	• ***	• Not N; approx. with N	A (rough)	---	---

a 5 is shown here as an arbitrary small # with reference to inappropriateness of RSS for non-additive systems when # of RVs is small.

* Functional relation Known

*** Functional relation Unknown

C" = Failure Rate Prediction method for which mathematical & computational techniques are related to C. C" can accommodate ≥ 1 failure condition.

@ A simple program for additive relationships; less computational time; applicable if only 1 failure condition

b While C also applies here, the purpose of approximating with normal distribution was to use B for reasons of enabling "hand" calculation or handling very large # of RVs that cannot be segmented into subsets for analyzing with C in a nested process with ≥ 2 steps (however, the latter is almost always doable).

NOTES: 1. C & C" apply to all cases where functional relation is known

2. 5th column not needed whenever CP is piece part dimension or material property only.

Abbreviations: PP = piece part; FR = Failure Rate; Mt'l Prop = Material Property completely covers the capability to quantify every link in the FOV chain for any type of distribution, any number of random variables, and any number of failure conditions or conditions that affect variability when functional relationships are known. The latter includes the case where the relationship is obtained experimentally and expressed as a matrix of discrete values, for this would either allow derivation of approximate functional relationships or using the values directly in the multiple integration scheme. (RSS should only be used for cases where the engineer or draftsman does not have any functional relationship or wants a gross estimate.)

WORK CONTENT AND INFRASTRUCTURE ENABLERS FOR TOLERANCING

While tolerance analysis, performance analysis, and tolerance synthesis can be carried out with assumed manufacturing tolerances and with inadequate cost information, it would be far more advantageous to know as much quantitative information about manufacturing and performance dissatisfiers in the Field as possible. For this, appropriate databases are needed.

Note that some of the databases are established through industry-specific internal testing, including that for material degradation and their effects relative to the industry's own applications; for example, elastomer/paper friction coefficient for Xerox. The construction of the databases drives a corporation's strategy for establishing infrastructure for system compatibility, information retrieval, vendor interaction, knowledge acquisition, etc. However, the most critical link of the UMDFL is to modeling, and it in turn drives much of the engineering content.

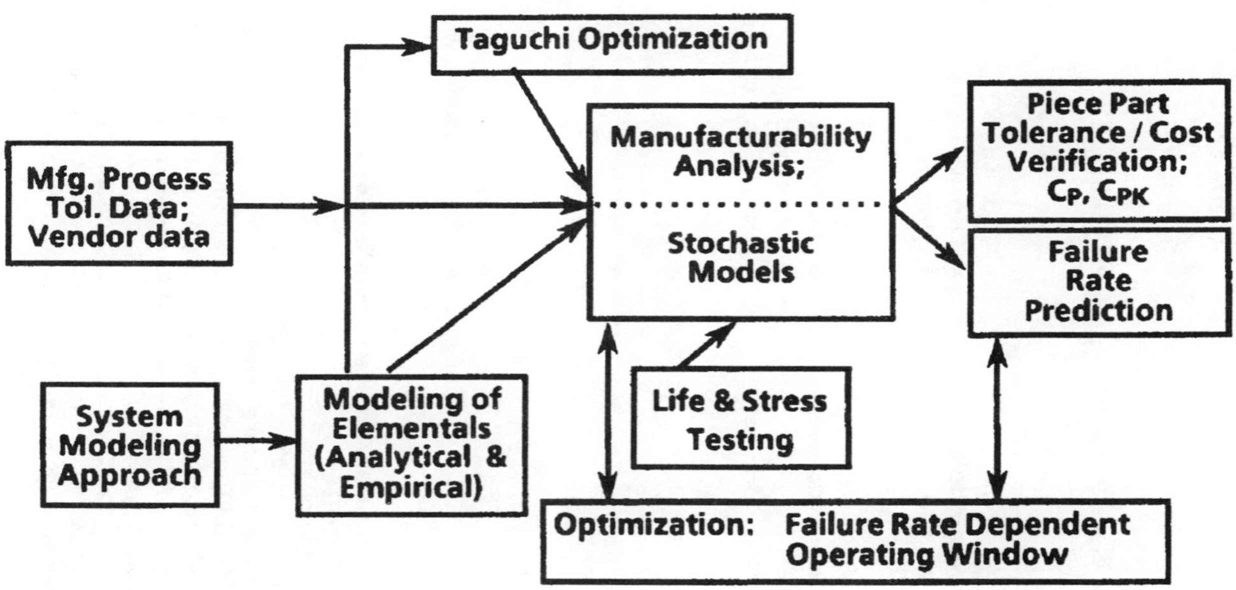

Figure 8 Product Synthesis and UMDFL: A Top Level View

Figure 8 shows a top level view of the major linked activities during product synthesis that relate to the UMDFL. The critical elements are an appropriate system modeling approach and the manufacturing databases. It also shows how Taguchi optimization and optimization through failure rate dependent operating window are linked. Manufacturability analysis involves the determination of whether or not the design intent can be manufactured (i.e., whether or not the required CPs and their specified tolerances could be met by the design concept generated, taking into account the part tolerances and the cost aspect). Thus, this includes the crucial step of verifying whether or not the manufacturing process capability (C_p, C_{pk}) is adequate and cost target is within range.

Note that optimization is a many step process; granularity differs. The failure rate dependent operating window optimization process allows for trade-offs between failure rates of different failure modes. For the next level of optimization, databases from the Field are needed. Thus, the UMDFL approach in fact unifies deterministic modeling with the stochastic modeling approach and theoretically establishes the linkage to the needed databases, knowledge bases, and the information retrieval/ management systems. The UMDFL provides a coherent view of the entire process.

To apply the UMDFL to most complex products does not necessarily involve modeling every physical relationship. One should selectively address those significant areas for which performance and cost trade-offs play major roles in design decisions. This makes modeling manageable for most of the chain. The modeling that relate lower levels of the FOV hierarchy to Critical Parameters variability is far less complex and should be carried out fully since it is the final part of tolerance analysis. Figure 9 puts in perspective the methods and tools of the UMDFL as well as the concept that the UMDFL drives a coherent design process.

RELATIONSHIP TO THE DESIGN PROCESS

Engineering and designing a product involves a process of reaching adequate understanding of how parts affect performance, and executing a design that has captured that understanding for optimizing the corporation's and the customer's values. A main question is how to attain quantitative understanding of the effect of variability and degradation of the parts on the product's performance. This would clearly involve mapping relationships for every hierarchical link from the most elementary levels to the performance level. Also, this would determine the necessary system modeling approach and the necessary first principles or empirical modeling activities. In essence, the information required for quantification at each step drives the necessary activities to supply that information. Thus, fleshing out the FOV chain defines a design process for quantitative engineering.

How much modeling should be done to yield enough understanding of variability propagation through the chain is intimately tied to a corporation's technology delivery process. The appropriate level of experimental and analytical work content for obtaining adequate models that provide relationships to appropriate random variables is dependent on the corporation's process capability. Analogy is made to implementation of Quality Function Deployment (QFD). For example, few corporations are capable of carrying out the full process of translating all elements of the voice of the customer down to the level of piece part specification. The most active corporation currently is implementing QFD for only a portion of the most important "voice" elements for any of its products and will include more over time. Similarly for the UMDFL, what should be modeled and how one should model depends on what are the important areas and what can be addressed within a planned time frame, while at the same time building according to a plan that would deliver enhanced and broader capabilities. Thus variability analysis and optimization can be carried out according to the granularity required at various stages of technology and product delivery processes and to the corporation's process capability.

A design process discussion is not complete without emphasizing the need for understanding how (1) to select Critical Parameters for which optimization is to be conducted to define their nominal values and tolerances and (2) to select parameters which should be considered as fixed but whose

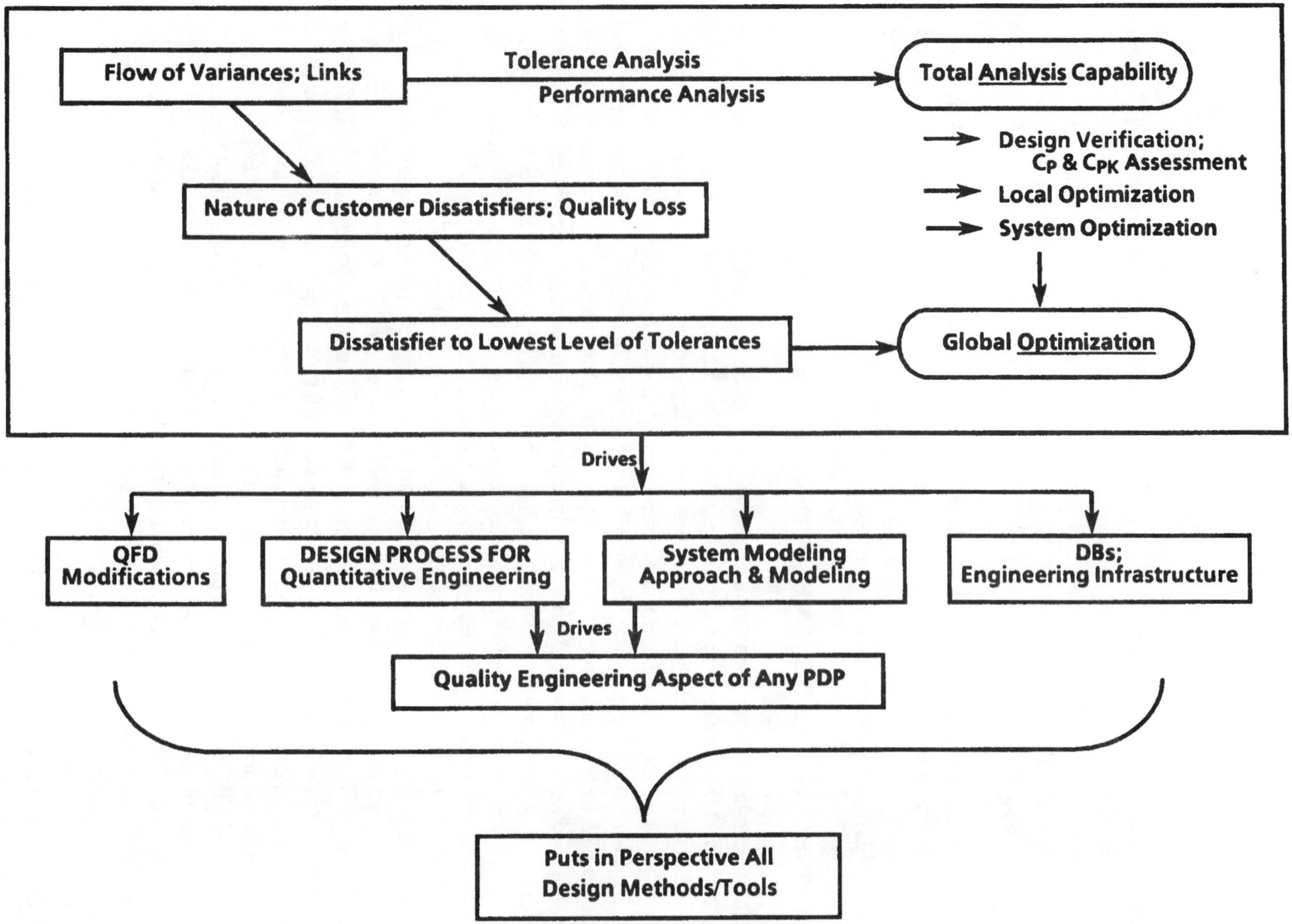

Figure 9: Summary of UMDFL concept

variability is to be considered as noise against which the set of optimized Critical Parameters would be made 'robust'. Such selection is an important part of the technology delivery process. Often much development effort must have taken place before this point, and much more yet needs to take place for optimization of the critical parameters. The importance of devising appropriate experiments that support development of models which can easily be converted to stochastic models with appropriate parameter variability cannot be overstated. These experiments and empirical models as well as analytical models are important both for reaching this selection point and beyond for optimizing the Critical Parameters. An often found deficiency in empirical models is that the experimental data is obtained from hardware with many assumed fixed parts. Subsequently, manufacturing variability may result in a design intent hardware with parts that differ from what were fixed parts in laboratory fixtures. It is no wonder that during product development, when manufactured parts are assembled into prototypes, performance shortfalls occur. It is therefore clear that a great deal of hierarchical modeling, both analytical and empirical, must take place to enable understanding the effect of variability along the FOV hierarchy.

A design process discussion is also incomplete without including one on Taguchi Methods. Dr. Taguchi's principles, Off-Line Quality Control, On-Line Quality Control, and use of the Quality Loss Function concept, for optimization and quality control certainly comprise the foundation of Quality Design. The Taguchi Methods, Parameter Design and Tolerance Design, should be an important part of technology and product delivery processes. These are most effectively employed when one has selected what should be the Critical Parameters, and are especially useful when models are too difficult to develop; the latter may be the situation for some complex subsystems and systems. For cases where one can model appropriately, one can use analytical methods for optimization or apply the Taguchi Method for calculating Signal-to-Noise (S/N) ratios after evaluating the models at the Orthogonal Array of set points. Having the models, however, one can use them for evaluating and optimizing beyond the Taguchi technique via the UMDFL approach because of the quantification capability. As we mentioned before, even for cases that cannot be fully modeled, and therefore, one must rely on Taguchi Methods to optimize, one can certainly model partially to determine the effect of variability at lower levels of the hierarchy and therefore support decisions on what should be Critical Parameters and what are their applicable ranges since the latter in part also depend on those other parameters that would be assumed to contribute the most to noise.

RESEARCH NEEDS IDENTIFIED

The development and application of the UMDFL has revealed a number of areas where additional research would significantly enhance capture of its full capability:

1. Fundamental Science and Modeling Methods: What Xerox did with paper handling elastomers, in working with our vendors to obtain formulations with more stable material properties and in doing internal experimentation at the fundamental level to understand the effect on elastomer-to-paper coefficient of friction due to paper types, velocity, pressure, relative humidity, degradation from our own machine contaminants, etc., is an example of this area of research need. This type of work, often industry specific, produces phenomenological relationships and variability information that feed into models for performance analysis. Additionally, general methods for determining phenomenological relationships are needed. Such relationships can provide guidance for empirical representations for practical applications.

2. Computational Techniques: Efficient computational techniques, such as significantly

improved quadrature and approximation techniques, are needed for dramatic reductions in computation time so that the UMDFL's tolerance synthesis capability can be more automated or that non-closed form relationships in the limits of integration can more generally be accommodated time-wise.

3. Databases for UMDFL implementation: A database structure for the UMDFL needs to be developed for efficient management of and rapid access to fundamental research/engineering, design, manufacturing, and service data, including their variability distributions. Linking this database with UMDFL and CAD tools should be accomplished for the engineers' convenient and efficient access.

ACKNOWLEDGMENT

The author owes gratitude to the following individuals for innumerable valuable discussions on design methodologies during the past 6 years: Maurice Holmes, Don Clausing, H. Barry Bebb, and Genichi Taguchi. Additionally, the author benefited from Bob Altavela's earlier work on implementation of the Quality Loss Function concept.

REFERENCES

Blanchard, Benjamin S., Fabrycky, Walter J. (1990), " Systems Engineering and Analysis", Prentice Hall, 1990.

Clausing, Don. P. (1984), " A comparison of Three Methods of Engineering Optimization", Tucson, AZ, 1984

Phadke, Madhav S. (1989), "Quality Engineering Using Robust Design", Printice Hall, 1989

Papoulis, Athanathios, "Probability, Random Variables, and Stochastic Processes", McGraw-Hill, 1965

Parks, Jean M. (1991), " Unified Methodology on Design for Latitude and the Role of Stochastic Modeling", Proceedings of the Design Productivity International Conference, Hawaii, February, 1991.

Parks, Jean M. (1980-82), Miscellaneous Xerox Internal Technical Reports and Memos, 1980-82.

Sorenson, Carl D., Nielson, Dale B., and Chase, Kenneth W. (1991), "Improved Methods for Tolerance Analysis of Mating Hole Patterns', Proceedings of the Design Productivity International Conference, Hawaii, February, 1991.

Tipnis, Vijay A., *Chairman*, "Research Needs and Technological Opportunities in Mechanical Tolerancing -- Results of An International Workshop," ASME CRTD-15, New York, 1989.

APPENDIX: APPLICATION OF UMDFL & SMB TECHNIQUE

While the SMB technique of UMDFL is applicable to a wide range of trade-off analysis involving performance/tolerance/cost, only one example is given below for brevity. It should be remarked that the SMB approach which the author developed (Parks, 1980-82) has been demonstrated to be applicable by the author on the problem of mated holes (Sorensson et al., 1991). To illustrate the application of the method, the following example from the discussion on the operating window for misfeed and multifeed is given.

OPERATING WINDOW OPTIMIZATION

Consider the one dimensional operating window of the paper feeder, for example, the operating range of the drive force, Z, bounded by limits associated with misfeed and multifeed rates. Let random variables X and Y represent customer noise, other critical parameters, or other noise parameters. Suppose the failure conditions for these two modes are given by

$$\xi_1(X,Z) = XZ < a \qquad \text{and} \qquad \xi_2(Z,Y) = ZY > b$$

where a and b are given constants. (In actual cases analyzed at Xerox, the failure condition for multifeed consists of a system of conditions) Keeping the common parameter Z as the argument in the probabilities of misfeed and multifeed, $P_1(Z)$ and $P_2(Z)$, and treating Z to be a deterministic variable at first we obtain,

$$P_1(Z) = \int_{-\infty}^{a/Z} f_X(X)dX \qquad \text{and} \qquad P_2(Z) = \int_{b/Z}^{\infty} f_Y(Y)dY$$

Thus $P_1(Z)$ and $P_2(Z)$ can be calculated when f_X and f_Y, the probability density functions are given (for example, X could be dependent on a set of customer noises or if X were a critical parameter, its nominal value and tolerance may have been predetermined). Then the failure rates,

$$\eta_1 = 10^6 P_1(Z) \text{...(for misfeed)} \qquad \text{and} \qquad \eta_2 = 10^6 P_2(Z) \text{...(for multifeed)}$$

in number of failures per million feeds behave as shown in the figure below.

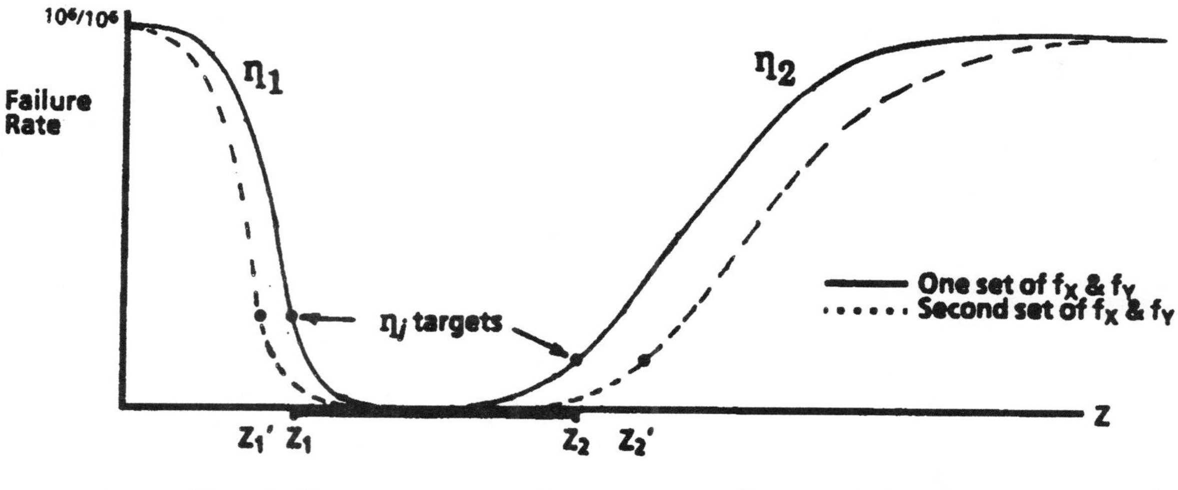

Operating Window of Z Associated with the Misfeed & Multifeed Rate Targets Is the Interval $Z = [Z_1, Z_2]$, or Widened to $[Z_1', Z_2']$ due to Changed f_X and f_Y.

Note that in this figure, the two failure rate curves (represented by dotted curves) change according to the distributions prescribed for X and Y. Thus , 'What-if' analysis on performance/ tolerance/cost trade-offs, including widening of the window or lowering of the cost with looser tolerance, can be conducted. Additionally, trade-offs between the failure rates of the feeder can be made (for example, instead of 50/million for each failure, 80/million plus 30/million, for misfeed and multifeed respectively, may yield appropriate customer satisfaction based on the trade-off of the operating window/tolerance/cost).

Next, consider Z to be stochastic as well. That is, Z will be the argument of $P_j(Z)$, where j=1&2, but the effect of statistical variability of Z itself (for example, Z may depend on friction coefficient and normal force which have variability) on $P_j(Z)$ is considered. $P_j(Z)$ then are given by

$$P_1(Z) = \int_{-\infty}^{\infty}\int_{-\infty}^{a/Z} f_X(X) f_Z(Z) dX dZ$$

$$P_2(Z) = \int_{-\infty}^{\infty}\int_{b/Z}^{\infty} f_X(Y) f_Z(Z) dX dZ$$

This is carried out by considering the argument Z in $P_j(Z)$ as the mean of Z while the distribution with this mean value of Z is used for evaluating the double integral. Rest of the process is covered in the text.

LIST OF PUBLICATIONS
ASME CENTER FOR RESEARCH AND TECHNOLOGY DEVELOPMENT

Volume No.	Title
CRTD-1	*Consensus of Current Practices for Lay Up of Industrial and Utility Boilers,* 1985, 28 pp. Book Number H00336
CRTD-2	*Research Needs in Thermal Systems,* 1986, 221 pp. Book Number I00212
CRTD-3	*Goals and Priorities for Research in Engineering Design,* 1986, 164 pp. Book Number I00230
CRTD-4	*Pressure Systems Energy Release Protection (Gas Pressurized Systems),* 1986, 425 pp. NASA Contractor Report 178090
CRTD-5	*Thermodynamic Data for Biomass Materials anmd Waste Components,* 1987, 496 pp. Book Number H00409
CTRD-6	*Analysis of the Feasibility of Replacing Asbestos in Automobile and Truck Brakes,* 1988, 136 pp. Book Number I00264
CRTD-7	*Hazardous Waste Incineration: A Resource Document,* 1988, 192 pp. Book Number I00266
CRTD-8	*Research Needs in Dynamic Systems and Control — Volume 1: Strategic Research Initiatives and Opportunities,* 1988, 58 pp. Book Number I00275
CRTD-9	*Research Needs in Dynamic Systems and Control — Volume 2: Acoustics and Noise Control,* 1988, 48 pp. Book Number I00276
CRTD-10	*Research Needs in Dynamic Systems and Control — Volume 3: Control of Mechanical Systems,* 1988, 40 pp. Book Number I00277
CRTD-11	*Research Needs in Dynamic Systems and Control — Volume 4: Machine Dynamics,* 1988, 37 pp. Book Number I00278
CRTD-12	*Research Needs in Dynamic Systems and Control — Volume 5: Nonlinear Dynamics,* 1988, 32 pp. Book Number I00279
CRTD-13	*Peer Review Thermal Energy Storage Program,* 1989, 46 pp. Available from CRTD, Washington, D.C.
CRTD-14	*The ASME Handbook on Water Technology for Thermal Power Systems,* 1989, 1900 pp. Book Number I00284
CRTD-15	*Research Needs and Technological Opportunities in Mechanical Tolerancing,* 1990, 51 pp. Book Number I00298
CRTD-15–1	*Selected Case Studies in the Use of Tolerance and Deviation Information During Design of Representative Industrial Products,* 1992, 156 pp. Book Number I00329
CRTD-16	*Engineering Fluid Mechanics Workshop Report,* 1990, 114 pp. Book Number I00299
CRTD-17	*ASME Research Committee on Corrosion and Deposits From Combustion Gases Seminar of Fireside Fouling Problems,* 1990, 183 pp.
CRTD-18	*ASME Ash Fusion Research Project,* 1990, 11 pp. Book Number I00307
CRTD-19	*Achievements in Tribology,* 1990, 173 pp. Book Number H00629
CRTD-20–1	*Risk-Based Inspection — Development of Guidelines: Volume 1, General Document,* 1991, 155 pp. Book Number I00311
CRTD-21	*Hazardous Waste Destruction by High Temperature Incineration,* 1991, 21 pp. Book Number I0266A
CRTD-22	*Mining Workshop for Nuclear Waste Cleanup,* 1991, 145 pp. Book Number I00314